Physical Science Gr
Table of Contents

Introduction	2
Curriculum Correlation	5
FOSS Correlation	5
Letter to Parents	6
The Scientific Method	7
The Science Fair	8
Assessments	9

Unit 1: What's the Matter?
- Background Information 14
- Describing Matter 18
- What's in the Bag? 19
- Solids, Liquids, and Gases 20
- Identifying Solids, Liquids, and Gases ... 22
- Finding Solids, Liquids, and Gases ... 23
- Measuring Solids 24
- Measuring Liquids 25
- Air Has Weight 26
- Molecule Detectors 27
- A Spice Chain 28
- Moving Molecules 29
- Water Facts 30
- From Solid to Liquid 31
- From Liquid to Gas 32
- How Long Did It Take? 33
- Taking Water Out of the Air 35
- From Water to Ice 36
- Heating Matter 37
- Cooling Matter 38
- Making Soap Bars 39
- Matter Changes Size 41
- Making Mixtures 42
- Making Solutions 43
- Matter in Water and Air 44
- Separating Mixtures by Freezing 45
- Separating Mixtures by Dissolving .. 47
- Matter in Air 49
- Matter in Water and Air 51
- Different Kinds of Change 52
- What Kind of Change? 53
- A Cooking Change 54
- A Model Fire Extinguisher 55
- Unit 1 Science Fair Ideas 57

Unit 2: The Force Is With You!
- Background Information 58
- Push or Pull? 64
- Measuring Forces 65
- More and More Force 66
- Calculating Force 67
- Gravity 69
- Is Gravity Working? 70
- Uphill and Downhill 71
- Which Takes More Force? 72
- The Force of Gravity 73
- What Can More Mass Do? 75
- What Causes Motion? 77
- Where Does It Go? 78
- Friction 79
- Friction and Lubricants 80
- Testing Friction 81
- Which Is the Easy One? 83
- Reaction Actions 84
- The Law of Motion 85
- Machines 86
- Types of Levers 87
- A Lifting Machine 89
- Pulleys 90
- Inclined Planes 91
- Wedges 93
- Screws 94
- Wheels and Axles 95
- Machines That Make a Difference ... 97
- Compound Machines 99
- Unit 2 Science Fair Ideas 100

Unit 3: The Energy of Life
- Background Information 101
- Blowing in the Wind 106
- Make a Dam 107
- Static Electricity 108
- What Is Electric Current? 110
- What Materials Conduct Electricity? ... 111
- Lemon Power 113
- Inside a Light Bulb 114
- Using Electricity 115
- How Can You Reduce Electrical Use in Your School? 116
- Safe or Not Safe? 117
- Baked Apple—The Solar Way 119
- Does the Heat from the Sun Affect Everything in the Same Way? 121
- Measuring Heat 122
- How Can You Measure Temperature? ... 123
- Measuring Heat 125
- Sources of Heat 126
- Popping Corn with a Candle 127
- Energy Detective 129
- Oil Traps 130
- Using Crude Oil 131
- Petroleum Around the World 133
- What Can a Magnet Pick Up? 135
- How Do Poles of a Magnet Act? 137
- A Magnet and an Electromagnet ... 139
- Unit 3 Science Fair Ideas 141

Answer Key 142

© Steck-Vaughn Company   Physical Science 3, SV 3762-3

Physical Science Grade 3
Introduction

We wake up in a new world every day. Our lives are caught in a whirlwind of change in which new wonders are constantly being discovered. Technology is carrying us headlong into the 21st century. How will our children keep pace? We must provide them with the tools necessary to go forth into the future. Those tools can be found in a sound science education. One guidepost to a good foundation in science is the National Science Education Standards. This book adheres to these standards.

Young children are interested by almost everything around them. They constantly ask questions about how and why things work. They should be encouraged to observe their world, the things in it, and how things interact. They should take note of the properties of the Earth and its materials, distinguish one material from another, and then try to develop their own explanation of why things are the way they are. A basic understanding of science boosts students' understanding of the world around them.

As children learn more about their world, they should be encouraged to notice changes that take place around them. These changes can be the evaporation of liquids into gas; the force needed to do work; how tools change force; and how energy is transferred to different materials to make new and different kinds of energy.

Organization
Physical Science serves as a handy companion to the regular science curriculum. It is broken into three units: What's the Matter?; The Force Is With You!; and The Energy of Life. Each unit contains concise background information on the unit's topics, as well as exercises and activities to reinforce students' knowledge and understanding of basic principles of science and the world around them.

- **What's the Matter?:** Matter is made up of three materials—solids, liquids, and gases. Students explore each of the states and how matter changes through the addition or reduction of heat. Physical and chemical changes are also identified. The unit also covers the basic principles of mixtures and solutions.
- **The Force Is With You!:** Students are introduced to the concepts of force and motion. The unit gives information on force, gravity, friction, and inertia and how they all interact to affect movement in the world. The unit then focuses on simple and compound machines. Students learn how pulleys, levers, inclined planes, screws, wedges, and wheels and axles make work easier.
- **The Energy of Life:** We flip a switch and a light comes on, but what makes that light work? The unit identifies a variety of energy sources, from wind and water to fuel. It then focuses on how the energy is transferred into electricity so we have light, heat, sound, and motion. Finally, students explore magnets and the force they produce.

This book contains three types of pages:
- Concise background information is provided for each unit. These pages are intended for the teacher's use or for helpers to read to the class.
- Exercises are included for use as tests or practice for the students. These pages are meant to be reproduced.
- Activity pages list the materials and steps necessary for students to complete a project. Questions for students to answer are also included on these pages as a type of performance assessment. As much as

possible, these activities include most of the multiple intelligences so students can use their strengths to achieve a well-balanced learning style. These pages are also meant for reproduction for use by students.

Use

Physical Science is designed for independent use by students who have been introduced to the skills and concepts described. This book is meant to supplement the regular science curriculum; it is not meant to replace it. Copies of the activities can be given to individuals, pairs of students, or small groups for completion. They may also be used as a center activity. If students are familiar with the content, the worksheets may also be used as homework.

To begin, determine the implementation that fits your students' needs and your classroom structure. The following plan suggests a format for this implementation.

1. Explain the purpose of the worksheets to your students. Let them know that these activities will be fun as well as helpful.
2. Review the mechanics of how you want the students to work with the activities. Do you want them to work in groups? Are the activities for homework?
3. Decide how you would like to use the assessments. They can be given before and after a unit to determine progress, or only after a unit to assess how well the concepts have been learned. Determine whether you will send the tests home or keep them in students' portfolios.
4. Introduce students to the process and the purpose of the activities. Go over the directions. Work with children when they have difficulty. Work only a few pages at a time to avoid pressure.
5. Do a practice activity together.

The Scientific Method

Students can be more productive if they have a simple procedure to use in their science work. The scientific method is such a procedure. It is detailed here, and a reproducible page for students is included on page 7.

1. PROBLEM: Identify a problem or question to investigate.
2. HYPOTHESIS: Tell what you think will be the result of your investigation or activity.
3. EXPERIMENTATION: Perform the investigation or activity.
4. OBSERVATION: Make observations, and take notes about what you observe.
5. CONCLUSION: Draw conclusions from what you have observed.
6. COMPARISON: Does your conclusion agree with your hypothesis? If so, you have shown that your hypothesis was correct. If not, you need to change your hypothesis.
7. PRESENTATION: Prepare a presentation or report to share your findings.
8. RESOURCES: Include a list of resources used. Students need to give credit to people or books they used to help them with their work.

Hands-On Experience

An understanding of science is best promoted by hands-on experience. *Physical Science* provides a wide variety of activities for students to do. But students also need real-life exposure to their world. Playgrounds, parks, and vacant lots are handy study sites to observe many of nature's forces.

It is essential that students be given sufficient concrete examples of scientific concepts. Appropriate manipulatives can be bought or made from common everyday objects. Most of the activity pages can be completed with materials easily accessible to the students. Manipulatives that can be used to reinforce scientific skills are recommended on several of the activity pages.

Science Fair
Knowledge without application is wasted effort. Students should be encouraged to participate in their school science fair. To help facilitate this, each unit in *Physical Science* ends with a page of science fair ideas and projects. Also, on page 8 is a chart that will help students to organize their science fair work.

To help students develop a viable project, you might consider these guidelines:
- Decide whether to do individual or group projects.
- Help students choose a topic that interests them and that is manageable. Make sure a project is appropriate for a student's grade level and ability. Otherwise, that student might become frustrated. This does not mean that you should discourage a student's scientific curiosity. However, some projects are just not appropriate. Be sure, too, that you are familiar with the school's science fair guidelines. Some schools, for example, do not allow glass or any electrical or flammable projects. An exhibit also is usually restricted to three or four feet of table space.
- Encourage students to develop questions and to talk about their questions in class.
- Help students to decide on one question or problem.
- Help students to design a logical process for developing the project. Stress that the acquisition of materials is an important part of the project. Some projects also require strict schedules, so students must be willing and able to carry through with the process.
- Remind students that the scientific method will help them to organize their thoughts and activities. Students should keep track of their resources used, whether they are people or print materials. Encourage students to use the Internet to do research on their project.

Additional Notes
- Parent Communication: Send the Letter to Parents home with students so that parents will know what to expect and how they can best help their child.
- Bulletin Board: Display completed work to show student progress.
- Portfolios: You may want your students to maintain a portfolio of their completed exercises and activities or of newspaper articles about current events in science. This portfolio can help you in performance assessment.
- Assessments: There are Assessments for each unit at the beginning of the book. You can use the tests as diagnostic tools by administering them before children begin the activities. After children have completed each unit, let them retake the unit test to see the progress they have made.
- Center Activities: Use the worksheets as a center activity to give students the opportunity to work cooperatively.
- Have fun: Working with these activities can be fun as well as meaningful for you and your students.

Physical Science Grade 3
Curriculum Correlation

Language Arts	18, 19, 38, 73, 81
Math	24, 25, 29, 30, 31, 33, 34, 36, 49, 50, 65, 67, 68, 82, 89, 108, 109, 122, 123, 124, 125
Science	20, 22, 26, 27, 32, 35, 37, 41, 43, 45, 46, 47, 48, 52, 57, 69, 72, 75, 76, 77, 78, 86, 87, 88, 90, 91, 92, 100, 110, 111, 112, 114, 115, 131, 135, 136, 137, 138, 139, 140, 141
Social Studies	53, 70, 73, 80, 81, 85, 93, 94, 99, 107, 116, 121, 126, 129, 130, 132, 133, 134
Health/PE	19, 21, 23, 28, 39, 40, 42, 44, 51, 54, 55, 64, 66, 71, 79, 83, 84, 95, 96, 97, 98, 106, 113, 117, 118, 119, 120, 127, 128

FOSS Correlation

The Full Option Science System™ (FOSS) was developed at the University of California at Berkeley. It is a coordinated science curriculum organized into four categories: Life Science; Physical Science; Earth Science; and Scientific Reasoning and Technology. Under each category are various modules that span two grade levels. The modules for this grade level are highlighted in the chart below.

Magnetism & Electricity	101-105, 108-109, 110, 111-112, 113, 114, 115, 116, 117-118, 129, 135-136, 137-138, 139-140, 141
Physics of Sound	See *Physical Science*, Grade 4.

Dear Parent,

During this school year, our class will be using an activity book to reinforce the science skills that we are learning. By working together, we can be sure that your child not only masters these science skills but also becomes confident in his or her abilities.

From time to time, I may send home activity sheets. To help your child, please consider the following suggestions:

* Provide a quiet place to work.
* Go over the directions together.
* Help your child to obtain any materials that might be needed.
* Encourage your child to do his or her best.
* Check the activity when it is complete.
* Discuss the basic science ideas associated with the activity.

Help your child maintain a positive attitude about the activities. Let your child know that each lesson provides an opportunity to have fun and to learn. Above all, enjoy this time you spend with your child. As your child's science skills develop, he or she will appreciate your support.

Thank you for your help.

Cordially,

Name _____ Date _____

The Scientific Method

Did you know you think and act like a scientist? You can prove it by following these steps when you have a problem. These steps are called the scientific method.

1. **Problem:** Identify a problem or question to investigate.
   _____

2. **Hypothesis:** Tell what you think will be the result of your investigation or activity.
   _____
   _____

3. **Experimentation:** Perform the investigation or activity.
   _____

4. **Observation:** Make observations, and take notes about what you observe.
   _____
   _____
   _____

5. **Conclusion:** Draw your conclusions from what you have observed.
   _____
   _____

6. **Comparison:** Does your conclusion agree with your hypothesis? If so, you have shown that your hypothesis was correct. If not, you need to change your hypothesis.
   _____

7. **Presentation:** Prepare a presentation or report to share your findings.
   _____

8. **Resources:** Include a list of resources used. You need to give credit to people or books you used to help you with your work.
   _____
   _____
   _____

Name _____ Date _____

The Science Fair

The science fair at your school is a good place to show your science skills and knowledge. Science fair projects can be several different types. You can do a demonstration, make a model, present a collection, or perform an experiment. You need to think about your project carefully so that it will show your best work. Use the scientific method to help you to organize your project. Here are some other things to consider:

Project Title _____			
Working Plan	**Date Due**	**Date Completed**	**Teacher Initials**
1. Select topic			
2. Explore resources			
3. Start notebook			
4. Form hypothesis			
5. Find materials			
6. Investigate			
7. Prepare results			
8. Prepare summary			
9. Plan your display			
10. Construct your display			
11. Complete notebook			
12. Prepare for judging			

Write a brief paragraph describing the hypothesis, materials, and procedures you will include in your exhibit.

Be sure to plan your project carefully. Get all the materials and resources you need beforehand. Also, a good presentation should have plenty of visual aids, so use pictures, graphs, charts, and other things to make your project easier to understand.

Be sure to follow all the rules for your school science fair. Also, be prepared for the judging part. The judges will look for a neat, creative, well-organized display. They will want to see a clear and thorough presentation of your data and resources. Finally, they will want to see that you understand your project and can tell them about it clearly and thoroughly. Good luck!

Name _____ Date _____

Unit 1 Assessment (Part 1)

Circle the best answer.

1. Matter keeps its own shape when it is in the form of
   - **a.** a solid.
   - **b.** a liquid.
   - **c.** a gas.
   - **d.** all of the above.

2. All of the following are the same form of matter except
   - **a.** milk.
   - **b.** lemon juice.
   - **c.** oil.
   - **d.** steam.

3. Moving gas from a jar to a balloon would change
   - **a.** the shape of the gas.
   - **b.** the gas to liquid.
   - **c.** the color of the gas.
   - **d.** nothing.

Use the drawing below to answer questions 4 and 5.

4. The water in the drawing is changing to
   - **a.** a solid.
   - **b.** a gas.
   - **c.** a liquid.
   - **d.** air.

5. The water in the drawing is
   - **a.** boiling.
   - **b.** condensing.
   - **c.** burning.
   - **d.** melting.

Name _____ Date _____

Unit 1 Assessment (Part 2)

6. Steel-wool pads rust faster when they are
  a. left outside in the rain.
  b. in a cardboard box on the grocery shelf.
  c. stored under the kitchen sink.
  d. kept in an airtight can.

7. Which picture shows that a solution has just been mixed?

 A   C

 B   D

  a. A   **b.** B   **c.** C   **d.** D

8. Your body needs the most oxygen when you are
  a. drawing a picture.   **c.** digesting your food.
  b. sleeping.   **d.** telling a joke.

9. Sugar would dissolve most easily in water that is at
  a. 0°C (32°F).   **c.** 65°C (150°F).
  b. 18°C (64°F).   **d.** 100°C (212°F).

Name _____ Date _____

Unit 2 Assessment

Circle the best answer.

1. Work is done when
   a. a force is used.
   b. a force moves an object.
   c. nothing moves.
   d. a person holds something.

Use the picture to answer questions 2, 3, and 4.

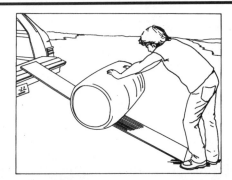

2. What machine is being used in the picture?
   a. inclined plane
   b. lever
   c. screw
   d. wedge

3. Another name for this machine is a
   a. spring scale.
   b. ramp.
   c. lever.
   d. wheel.

4. This machine makes work easier because it
   a. has a sharp point.
   b. has a slanted surface.
   c. moves slowly.
   d. moves quickly.

5. An inclined plane that winds around in a spiral is a(n)
   a. lever.
   b. wedge.
   c. screw.
   d. ax.

6. A spring scale is used to measure
   a. speed.
   b. distance.
   c. force.
   d. time.

Name _____ Date _____

Unit 3 Assessment (Part 1)

Circle the best answer.

1. Objects with static electricity can pick up
   a. paper.   **c.** iron.
   b. heavy objects. **d.** magnets.

2. If a circuit is open,
   a. a light will go on.
   b. all the parts are connected.
   c. the switch may be off.
   d. a current will go through it.

3. Current moves easily through
   a. an insulator.   **c.** a charge.
   b. a conductor.   **d.** an open switch.

4. The source of electricity in a flashlight is the
   a. switch.   **b.** metal.   **c.** dry cell.   **d.** bulb.

5. Which of these drawings shows a closed circuit?

   a. A   **b.** B   **c.** C   **d.** D

Unit 3 Assessment

Unit 3 Assessment (Part 2)

6. A magnet will pick up an object that is made of
  a. plastic.   **c.** wood.
  b. iron.   **d.** copper.

7. Which of the following can magnets pick up?
  a. paper clips   **c.** pennies
  b. rubber bands   **d.** small pieces of paper

8. A bar magnet would pick up pins
  a. only at its north pole.
  b. at all points equally.
  c. mainly at its north and south poles.
  d. only in its middle.

9. A magnet will pick up a
  a. plastic fork.   **c.** wooden bowl.
  b. steel knife.   **d.** coin.

Use the drawings to answer question 10.

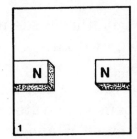 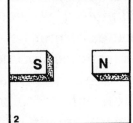 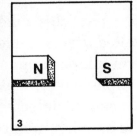 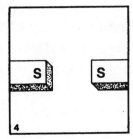

10. In which two drawings will the poles pull at each other?
  a. 1 and 2   **b.** 2 and 3   **c.** 1 and 4   **d.** 3 and 4

Unit 1 | What's the Matter?

INTRODUCTION

By the third grade, students have become competent in using the basic skills of reading, writing, and simple calculations. They have progressed from a world of concretes to an interest in understanding their world through exploration. They do not always have to see to believe. Students begin to question the whys and the hows of everything around them. This unit introduces the basic component of physical science first taught—matter. Students explore the three states of matter and how different kinds of changes affect each state in an activity-based format.

MATTER

Matter is all around. It is everything that we see and touch. Moreover, matter has mass, or weight, and takes up space. Matter is identified in three forms—solid, liquid, and gas. While students can easily comprehend and recognize the properties of a solid and liquid, it is generally difficult for them to understand that air is matter. Students begin to grasp this complex concept as they experiment with balloons and candles. (Energy is an example of something that is not matter and will be explored in Unit 3.)

Matter can be easily described by its properties, both physical and chemical. Physical properties describe how a substance looks, which includes color, shape, texture, melting point, and boiling point. By using their senses, students can describe what an item looks and feels like. Chemical properties tell how something reacts with another substance so that it changes in its appearance, taste, or smell. For example, iron reacts with oxygen and water to make a new substance—rust.

All matter is made up of tiny particles called *molecules*. Molecules are made up of even smaller particles called *atoms*. Molecules cannot be seen with a microscope, but students can understand a substance's properties by using their senses when performing simple experiments. If sugar is dissolved in water, the sugar cannot be seen; but it can be detected through taste because the water is sweet. By using the sense of smell, students can identify molecules of vinegar in air, a gas. To some degree, hearing can be used to sense molecules, because a smoke detector detects molecules of smoke in the air and buzzes to alert people to the potential danger.

SOLIDS

The state of matter is determined by the density of the molecules and how fast they move. In a solid, the molecules are attracted to each other and are tightly held together. The movement of the particles is limited—they vibrate only. Therefore, a solid has a definite shape and volume. For example, a rock has a certain shape. It can be broken into smaller pieces, but its molecules do not change. A solid's mass is measured in grams (g), a metric weight which is a scientific measurement standard.

LIQUIDS

Liquids have a definite volume, but they take the shape of the container. The molecules in a liquid are not packed as tightly, so can move about more freely and easily by sliding over each other. This movement is what makes a liquid take the shape of the container. When juice is in a carton, it takes the shape of the carton. Yet if poured into a glass, the juice takes the shape of the glass. The volume of a liquid is measured in milliliters (mL), the scientific standard measurement for liquid.

GAS

Gas is the third state of matter. It is harder for students to understand the properties of gas, because they cannot see it, nor have they had exposure to different kinds of gases. In a gas, the molecules are far apart and move very quickly and randomly in all directions. They bounce off of each other when they collide. Gas has no definite shape or volume. Gas, therefore, expands to take the shape of a container. Gas is also measured in milliliters (mL).

MATTER CHANGES

All matter can change form, meaning it can change from one state to another. When matter changes, nothing is lost or gained—the molecules stay the same. The addition or the removal of heat causes the molecules to get closer or farther apart. Moreover, the greater the amount of heat, the faster the molecules move. These changes in the density and the speed of a substance's molecules cause the state of matter to change.

When a solid is heated, the molecules expand. The heat causes the speed and volume of molecules to change. They vibrate faster and slip out of position, resulting in the solid changing into a liquid. This process is called *melting,* and the point at which the solid changes to a liquid is called the *melting point*. All matter, including rocks, has a melting point. The most commonly recognized melting point (or freezing point) is that of water, which is 0° on the Celsius scale or 32° on the Fahrenheit scale. Even with this change, the structure of the molecules stays the same.

When liquid is heated, the loose molecules continue to expand. The vibration increases, causing them to collide with each other and move in all directions. When the *boiling point* is reached, the liquid changes into a gas. The most commonly recognized boiling point is that of water. It boils at 100° Celsius or 212° Fahrenheit. This process is called *evaporation*. Again, the molecules stay the same; nothing is lost or gained when the matter changes states.

The removal of heat causes the reverse changes in matter. Through *condensation*, a gas is cooled, and the molecules contract. They stop colliding and return to their loosely packed state, thus becoming a liquid. If heat is removed to the point that a liquid reaches its freezing point, a liquid will become a solid. The molecules are densely packed and cannot slide around. In any of these changes, nothing is lost or gained; only the properties of matter change.

Students can easily experiment with changes in states by watching ice change to water and steam. Ice is a solid, but when heated to its freezing point, turns to liquid water. No water is lost or gained in the process, and no molecules are changed. When more heat is added, the water changes to a gas called water vapor when it reaches its boiling point. The gas cannot be seen, because it has no color. Again, no water is lost or gained, and the molecules stay the same. If a spoon is held in the water vapor, the surface temperature of the spoon, which is room temperature, causes the water vapor to cool and condense back to liquid water. Likewise, by removing the heat and freezing the water, it changes states again to become ice.

Physical Changes
Matter can be changed in two ways, either in a physical change or in a chemical change. A physical change in matter is a change in which the molecules of a substance or substances do not change. There are three kinds of physical change. When matter changes states, as explained above, it is one kind of physical change.

A second kind of physical change takes place when a mixture is made. A *mixture* is a combination of substances in which the molecules of the substances diffuse evenly. Each substance retains its own properties and can be detected by the senses. None of the molecules is lost, gained, or changed. Moreover, a mixture can be separated by physical means, such as filtering, sorting, heating, or evaporating.

Each state of matter can make a kind of mixture. A fruit salad is an example of a solid mixture easily explained to students. Students can see and taste each piece of fruit. Sorting the fruit chunks is possible. Gas can also be mixed with another substance to form a mixture. The scent of a flower mixes with air so that you can smell the flower several feet away. You use your sense of smell to become aware of the scent. A liquid can also be a made into a mixture. Often a solid is dissolved into a liquid. This kind of mixture is called a solution. It is hard to separate out the parts, but it can be done. Lemonade is a good example to explain a solution. Water, lemon, and sugar are mixed together. Even though it does not look like a mixture, the ingredients can be separated. The lemon can be filtered out. The water can evaporate, leaving crystals of sugar.

A third kind of physical change takes place when the shape of a substance is changed through cutting, ripping, or grinding. A log can be cut into many pieces. What remains are sawdust and cut logs. The molecules of the log itself have not changed.

Chemical Changes
When the molecules of a substance change, a chemical change has taken place. A new substance is always made in a chemical change, but molecules are never lost. Even though new molecules are made, the same number of atoms exists. Energy, generally in the form of heat, causes the atoms in molecules to break down to form different molecules. Baking is a common example of chemical change. Sugar, milk, eggs, and flour are combined to make a cake batter mixture. When heat is added, a chemical change takes place to turn the ingredients into a cake. Chemical changes also occur in the human body. Through chemical changes, food and oxygen react in the body's cells to create energy to make the body work.

Name _____   Date _____

Describing Matter

Everything you see, taste, touch, and smell is made of matter. Matter is anything that has weight and takes up space. A rock is made of matter, and even air is made of matter!

You will need:
4 or 5 kinds of things
pencil

Matter Description

1. Write the name of your items at the top of each column.

2. Look at the color of each thing. Look at the shape. Feel each one. Can you smell anything?

3. On the table, make a list of words you could use to describe each thing. Use words like *soft, smooth, round,* and *lemony.*

4. Be sure to use color words. Also, tell what each thing is made of.

Answer the questions.

1. What is everything made of? _____

2. Are any two things described the same way? Explain.

_____

_____

Name _____ Date _____

What's in the Bag?

This is a guessing game you can play with two or more friends.

You will need:
a paper bag
4 or 5 "mystery" objects

1. Find objects that will fit in the bag. Make sure your friends don't see what they are. Put one object in the bag.

2. Have one friend put a hand inside the bag and feel the mystery object. Make sure your friend doesn't look at it. Have your friend describe the object using words like *hard, soft, round, square, light, heavy, smooth, rough*.

3. Ask your other friends to guess what the mystery object is.

4. Take turns placing objects in the bag, describing them, and identifying them.

Answer these questions.

1. What are all the objects in the bag made of? _____

2. How do you find out about matter? _____
_____

Unit 1: What's the Matter?

Name _____ Date _____

Solids, Liquids, and Gases

Everything around you is matter—the chair you are sitting on, the air around you. You are matter, too. All matter has some things in common. For instance, all matter takes up space. Matter is commonly found on Earth in one of three states—solid, liquid, and gas. Matter in each of these states has different properties.

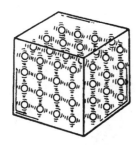  

A *solid* has a definite shape. It also has a definite volume. That is because its particles are very close together and are in a regular pattern. The particles move within the solid, but they are held together by an attraction. Heating a solid causes its particles to move more rapidly, weakens the attraction between them, and melting occurs.

A *liquid* does not have a definite shape. However, it does have a definite volume. The particles that make up a liquid move more rapidly and freely than those in a solid. The attraction between them is not as strong as the attraction between the particles in a solid, and the particles tumble over and around each other. A liquid flows and takes the shape of the container into which it is poured.

A *gas* does not have a definite shape or volume. The particles that make up a gas move rapidly and freely. They don't have much attraction for one another. A gas spreads out to fill its container. The air around you is a gas.

GO ON TO THE NEXT PAGE ▶

Name _____ Date _____

Solids, Liquids, and Gases, p. 2

Think about the difference between solids, liquids, and gases. Then fill in the blanks in the chart below.

STUDYING MATTER

	Solid	Liquid	Gas
Does it take up space?			
Does it have a shape of its own?			
Does its shape depend on the shape of its container?			
Can it be seen?			
Does it always stay the same size?			
Does it spread out to fill up its container?			

Answer the questions.

1. What are the three states of matter?

   _____

2. Look around the room. In what state are most objects? List three of them.

   _____

   _____

   _____

3. Identify something in your body that is a solid, something that is a liquid, and something that is a gas.

   _____

   _____

   _____

Name _____   Date _____

Identifying Solids, Liquids, and Gases

Each of these pictures shows matter that is a solid, a liquid, or a gas. Write the form of matter under each picture.

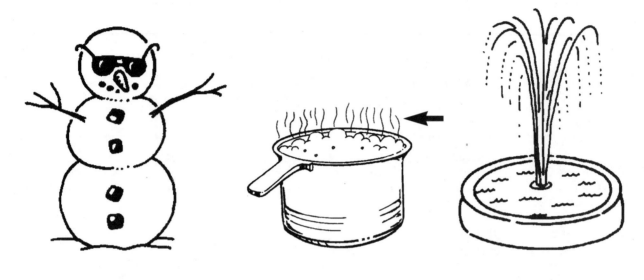

1. _____   2. _____   3. _____

4. _____   5. _____   6. _____

Unit 1: What's the Matter?

Name _____ Date _____

Finding Solids, Liquids, and Gases

All things are made of matter. Matter takes up space. It comes in different forms. It may be a solid, like wood. It may be a liquid, like milk. It may be a gas, like air.

You can find out about matter by using your senses. Look for examples of each form of matter. Let your senses of sight, touch, smell, taste, and hearing help you.

 Write your examples of matter in the list below. Then tell if each is a solid, liquid, or gas. Put an X in the correct column.

Kinds of Matter

	Name of Matter	Solid	Liquid	Gas
1.	Example: rock	X		
2.				
3.				
4.				
5.				
6.				
7.				
8.				
9.				
10.				

Unit 1: What's the Matter?

Name _____ Date _____

Measuring Solids

A balance is used to measure a solid object's mass, or weight. Balances have two pans. In one pan you place the object you want to measure. In the other you place the gram (g) weights, the metric units used by scientists to measure mass.

 You will need:
balance   gram weights   4 or 5 objects

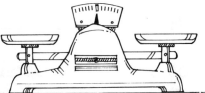

1. Make sure the empty pans are balanced. They are in balance if the pointer is at the middle mark on the base. If the pointer is not at this mark, move the slider to the right or left.

2. Place one object in a pan. The pointer will move toward the pan that is empty.

3. Choose weights to place in the other pan. The pointer will move back toward the middle. When the pointer is in the middle, the pans are balanced. Add the numbers on the gram weights. The total is the mass of the object.

4. Fill in the chart as you go.

5. Repeat with the other objects.

Measuring Solids

Object	Mass (g)

 Answer the questions.

1. Why must the pointer be in the middle before you begin? _____

2. How do you find the mass of an object once the balance is equal? _____

3. List your items in order from the least mass to the most mass.

_____

Unit 1: What's the Matter?

Name _____ Date _____

Measuring Liquids

Containers for measuring liquids are made of clear or translucent materials so that you can see the liquids inside them. On the outside of each of these measuring tools, you will see lines and numbers that make up a scale. On most of the containers used by scientists, the scale is in milliliters (mL).

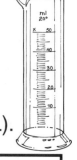

 You will need:
water   graduated cylinder   beaker   3 or 4 small containers

1. Pour water in each container until the containers are full.

2. Set the cylinder and beaker on a flat table, with scales facing you.

3. Pour the water from 1 container into the cylinder or beaker. (For large amounts, use the beaker.) Move so your eyes are even with the surface of water in the cylinder.

4. Find the scale line that is even with the top of the liquid. The surface may look curved. Take your reading at the lowest point of the curve. You may have to estimate the volume of the liquid if the surface falls between the numbers on the scale. Decide which line the water is closer to and use that number.

5. Fill in the chart as you go.

6. Repeat with the water in the other containers.

Measuring Liquids

Object	Milliliters (mL)

 Answer the questions.

1. What state of matter does a graduated cylinder measure? Why?
_____

2. Would you use a beaker or a cylinder to measure a can of juice?
_____

3. List your water amounts in order from the least milliliters to the most milliliters. _____

Unit 1: What's the Matter?

Name _____   Date _____

Air Has Weight

An empty bottle isn't really empty. An empty bag isn't empty, either. Inside each container is a colorless, odorless, gas–air. Here is how you can show that air has weight.

You will need:
2 balloons of the same size   meter stick
3 pieces of string   wire coat hanger   pin

1. Blow up the balloons to the same size. Tie the ends with strings so that the air cannot escape. Leave long pieces of string attached to each balloon.

2. Tie a balloon to each end of the meter stick.

3. Tie the third string around the middle of the meter stick. Hang it so it swings freely. Adjust the string until the meter stick balances.

4. When the stick is still, ask an adult to pop one of the balloons with a pin.

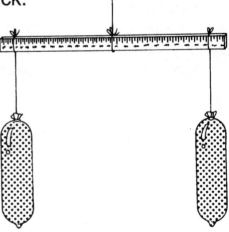

Answer the questions.

1. What happened when one balloon was popped? _____
_____

2. Why did the balloon sink? _____
_____

3. What makes the difference in weight? _____
_____

Unit 1: What's the Matter?

Name _____ Date _____

Molecule Detectors

Molecules are small particles in matter that you cannot see. Your senses can often tell you when molecules travel in the air or through a liquid. For example, you can smell hot soup when molecules of the soup reach your nose. When you add food coloring to water, you see the color as it moves through the water.

Look at the pictures below. In each one something is happening that causes molecules to travel through air or a liquid. On the first line under each picture, write which sense would tell you that molecules moved from one place to another. On the second line write *air* or *liquid* to tell where molecules have moved.

1.
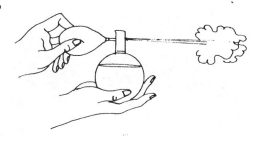

_____

_____

2.

_____

_____

3.
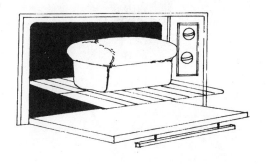

_____

_____

4.

_____

_____

Unit 1: What's the Matter?

Name _____ Date _____

A Spice Chain

Cooks use spices in foods because they add flavor. They have good odors, too. Some spices can be used to make unusual neck chains, too. When you wear the chain, you can smell the spices.

✓ **You will need:**
whole cloves   small bowl   nylon thread
whole allspice   scissors   water   large needle

1. Put the cloves and allspice in a bowl. Soak them in water overnight.

2. Cut a long piece of thread. Thread it through the needle. Tie a knot at the far end of the thread.

3. Use the spices in the same way you would use beads. With your teacher's help, string them in an interesting pattern. When you are finished, tie a knot in the loose end of the thread. Tie both ends of the chain together.

4. Let the chain dry for a day or two. The spices will shrink back to their original shapes. Your chain is ready to wear.

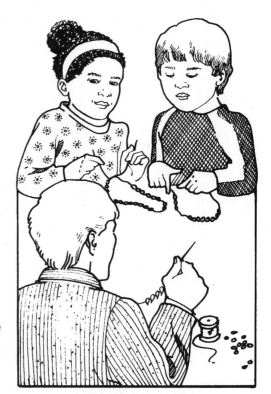

Answer the questions.

1. How does the chain smell? _____

2. What things smell good to you? _____

3. Why do you smell things? _____

Unit 1: What's the Matter?

Name _____ Date _____

Moving Molecules

A. Molecules are so tiny that they can pass through the tiny holes in the fabric of a balloon.

✔ **You will need:**
a balloon

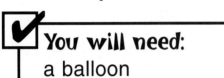

1. Blow up the balloon and tie the neck tightly. Set it aside for a week.

2. Look at the balloon again.

   1. How has the balloon changed? Why? _____
   _____
   _____

B. Molecules in liquids move from place to place. Find out whether molecules move faster in warm water or cold water.

✔ **You will need:**
2 clear plastic cups   food coloring
watch with a second hand   water

1. Fill one cup with very warm water.
Fill the other with cold water.

2. Put two drops of food coloring in each cup at the same time. (You will need to have a friend help you.) Write your starting time on the chart. Then time how long it takes for the food coloring to spread in the cups.

	Warm Water	Cold Water
Starting Time		
Finishing Time		

   2. Did it take longer for the color to spread in the warm or cold water? Why? _____
   _____

Unit 1: What's the Matter?

Water Facts

Water has no odor or taste. Below 0°C, it is solid ice. It is hard and cold, and it can float on liquid water. From 0°C to 100°C, water is a liquid. It pours and flows. Above 100°C, water is a gas called water vapor. Water vapor cannot be seen.

In its solid and liquid forms, water covers about 70 percent of the surface of the Earth. Water is in the air, too. Much of the time it is invisible water vapor. Often, however, water vapor in the air changes form. Then you can see it as clouds, rain, fog, snow, hail, or sleet, or as dew or frost on plants and cars.

Most of your body is water. Water makes up 92 percent of the liquid part of your blood. It makes up 80 percent of your muscles. You need water to live. A person can live without water for only 7 to 10 days.

Answer the questions. Write *true* in the space if the statement is correct. Write *false* if the statement is wrong.

_____ 1. Ice floats on liquid water.

_____ 2. Water forms a gas at 0°C.

_____ 3. Water vapor in the air may become frost.

_____ 4. You can see water vapor.

_____ 5. A person can live without water for a month.

Name _____ Date _____

From Solid to Liquid

Matter can change states. Solids can become liquids, and liquids can change to solids. How does this happen? Heat is either added to or removed from matter to cause state changes. Think about an ice cube. If you hold it in your hand, it gets warmer and begins to melt (becomes a liquid). However, if you put water in a freezer, it freezes (becomes a solid).

You will need:
ice cubes
small self-sealing plastic bag
watch or clock

1. Think of ways to make ice cubes melt quickly. Choose the idea that you think is best.

2. Place an ice cube in the bag. Try your idea. Use a watch or clock to time it.

3. Write the time it took to melt the ice cube. Compare your time with the times of some of your classmates.

Answer the questions.

1. How long did it take for your ice cube to melt?

_____

2. What made the ice cube melt? _____

3. What was the quickest way you found to melt an ice cube?

_____

_____

Unit 1: What's the Matter?

Name _____ Date _____

From Liquid to Gas

Remember, matter can change states. Think about what happens when you climb out of a swimming pool on a hot day. You leave wet footprints as you walk across the cement. The liquid water evaporates to become water vapor in the air. If you cool the air, water vapor condenses on objects such as grass and leaves. In this experiment, you will see how heat makes liquids evaporate.

You will need:
2 paper towels
1 glass of water

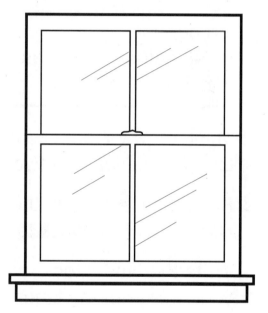

1. Sprinkle water on the paper towels. Do *not* put a lot of water on them.

2. Place 1 towel in a cool, dark place. Place the other towel on a sunny windowsill.

3. Every few minutes, check to see if the towels are dry.

Answer the questions.

1. Which towel was dry first? _____

_____

2. Why did this towel dry first? _____

_____

3. Where did the water in the towel go? _____

_____

Unit 1: What's the Matter?

Name _____ Date _____

How Long Did It Take?

If you leave a cup of water out long enough, the water will all evaporate. What happens if you pour the same amount of water into a pie tin or a soda bottle? Does it take the same amount of time to evaporate? Does it take more or less time?

✓ **You will need:**
large paper cup	measuring cup	funnel
pie tin	soda bottle	water

1. Put the paper cup, the pie tin, and the soda bottle on a table where they can stay for a week. Make sure they are out of direct sunlight and away from drafts. Fill the measuring cup with water. The cup holds 250 milliliters. Pour the water into the paper cup.

2. Pour one measuring cup of water into the pie tin. Using the funnel, pour one measuring cup into the soda bottle.

3. The next day, carefully pour the water from the paper cup into the measuring cup. Record the water level in the graph. Pour the water back into the paper cup.

4. Do the same thing with the water in the soda bottle. Use the funnel when you pour the water back into the bottle.

5. Ask your teacher to help you pour the water from the pie tin into the measuring cup. Record your measurements in the graph.

6. Every day for a week, measure the water left in each container. Record your results.

GO ON TO THE NEXT PAGE ➤

© Steck-Vaughn Company

33

Unit 1: What's the Matter?
Physical Science 3, SV 3762-3

Name _____ Date _____

How Long Did It Take?, p. 2

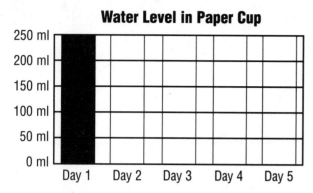

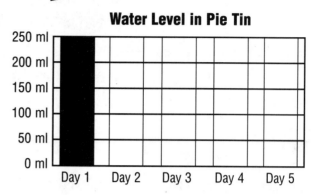

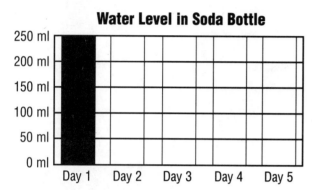

Answer the questions.

1. From which container did the water evaporate most quickly? _____

    How long did it take? _____

2. From which container did the water evaporate most slowly? _____

    How long did it take? _____

3. How can you explain your results? _____

    _____

    _____

Unit 1: What's the Matter?

Taking Water Out of the Air

Putting water into the air is something you have done often. For example, when you wash your hair, the water doesn't stay on your hair. It evaporates. But have you ever taken the water out of the air? Here is a way to change water vapor to liquid water.

✓ **You will need:**

| empty metal can | water | 3 ice cubes |
| paper towel | spoon | food coloring |

1. Fill the can halfway with cold water. Put in three drops of food coloring and stir.

2. Add the ice cubes. Wipe the outside of the can with the paper towel. Make sure the can is dry. Wait a few minutes.

 Answer the questions.

1. What forms on the outside of the can? _____

2. What color are the drops? _____

3. Did they come from inside the can? _____

4. How do you know? _____

The drops came from water vapor in the air. When water vapor is cooled, it collects into water droplets.

5. Where else have you seen water droplets collect? _____

Unit 1: What's the Matter?

Name _____ Date _____

From Water to Ice

When water turns to ice, it changes state. It changes from a liquid to a solid. Does the weight of the water change, too?

✓ **You will need:**
balloon   water   scale   freezer

1. Over a sink or basin, fill the balloon with water. Tie the open end into a tight knot.

2. Place the water-filled balloon on a scale. What does it weigh? _____

3. Put the water-filled balloon in the freezer. Leave it overnight. When all the water in the balloon has turned to ice, place the balloon on the scale.
What does it weigh? _____

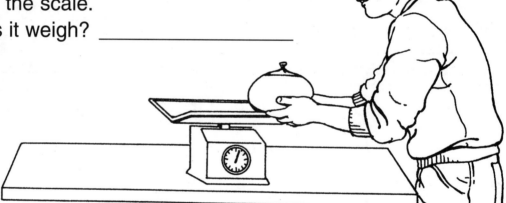

 Answer the questions.

1. The water in the balloon changed from liquid to solid form. Did the weight of the water change? _____

2. How do you explain what happened? _____
_____
_____

© Steck-Vaughn Company

Unit 1: What's the Matter?
Physical Science 3, SV 3762-3

Heating Matter

When matter is heated, it can change states. Solids change to liquids. Liquids change to gas. Gas expands, or gets bigger.

1. In the box at the right, draw how the glass would look if the liquid in it evaporated.

2. In the box at the right, draw how the balloon would look if the air in it were heated.

3. In the box at the right, draw what will happen if the ice cube is not placed in the freezer.

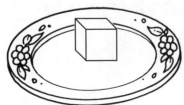

Unit 1: What's the Matter?

Name _____   Date _____

Cooling Matter

One thing can cause something else to happen. The thing that happens is called an effect. Read the lists of causes and effects. Then write the missing causes or effects in the blank spaces.

Cause **Effect**

1. _____   Water condenses on the outside of a glass of ice water.
   _____
   _____
   _____

2. The freezer cools a tray of water to 0° Celsius.   _____
   _____
   _____
   _____
   _____

3. Hot liquid iron is poured into a container and left to cool.   _____
   _____
   _____
   _____
   _____

4. _____   Steam condenses into liquid water.
   _____
   _____
   _____

Unit 1: What's the Matter?

Name _____ Date _____

Making Soap Bars

When heat is applied to some solids, they turn to liquid. When cooled, they change back to the same solid state. Melting solids and reusing them is a good way to recycle many things. Glass and metals, such as aluminum, are often reused in this way. Small pieces of soap can be recycled, too. They are too small to use for washing. It seems wasteful to throw them away. There *is* something you can do. Recycle them.

✔ **You will need:**

small pieces of handsoap	pot holders
small cooking pot	wooden spoon
wax paper	knife
cookie cutters	perfume (optional)
water	heat source

NOTE: This experiment must be done with an adult.

1. Ask your teacher to cut the soap into small pieces. Place the soap pieces in the pan.

2. Add just enough water to cover the soap. Put the pan aside for several hours until the soap becomes soft. When it is ready, it will look like pudding.

3. Add a few drops of perfume to the mixture to give it a pleasant odor. Stir.

4. Ask your teacher to help you cook your soap. Place the pan on a hot plate. Cook the soap over low heat. Stir it as it cooks. When the mixture melts and becomes smooth, ask your teacher to remove the pan from the heat.

GO ON TO THE NEXT PAGE ▶

Unit 1: What's the Matter?

Name _____   Date _____

Make Soap Bars, p. 2

5. Keep stirring the soap while it cools. You can stop stirring when it feels like soft dough.

6. Remove the soap from the pan. Put it on a piece of wax paper. Flatten it with your hands.

7. Using cookie cutters, cut the soap into different shapes. Put the finished soap bars aside. Let them harden overnight.

Answer the questions.

1. What states of matter did the soap form?
   _____

2. What made the soap change states?
   _____

3. How did you recycle the soap?
   _____

4. Why is it important to recycle?
   _____
   _____

Unit 1: What's the Matter?

Matter Changes Size

When water freezes, it expands. But most gases, solids, and liquids contract when they are cooled. If you heated a balloon, it would expand. A sidewalk would contract if it were cooled.

1. This series of pictures shows a balloon that is being heated. Number the pictures from warmest to coolest (1, 2, 3).

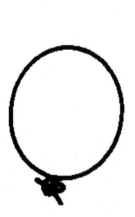  

_____   _____   _____

2. In these pictures, one jar lid is being cooled. The other jar lid is being heated. Circle the picture that shows the lid that would be easier to take off.

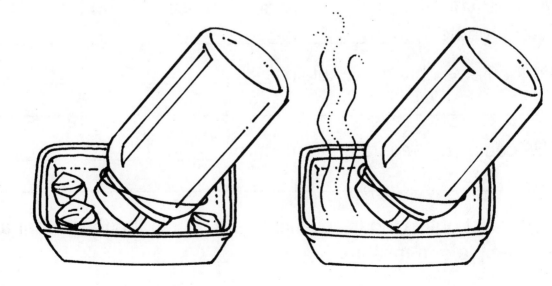

Name _____ Date _____

Making Mixtures

A mixture is made when two or more items are mixed together, but each keeps its same properties. For example, a salad is a mixture. You can see the lettuce, carrots, celery, and salad dressing. The foods have not changed—they look and taste the same. Moreover, a mixture can also be separated.

You will need:
sand   marbles
water   clear container
spoon

1. Pour some sand into the jar. Add several marbles. Mix them with the spoon.

2. Look at the container to see if the sand or marbles have changed.

3. Pour water into the container. Mix it all together. Look to see how all three items have changed.

Answer the questions.

1. What did you make when you added the sand, marbles, and water together? _____

2. Did the items change? Explain. _____
_____

3. How could you separate the mixture? _____
_____

4. Make a list of five foods you like to eat. Circle the foods that are an example of a mixture. _____

Unit 1: What's the Matter?

Name _____ Date _____

Making Solutions

A solution is a kind of mixture. It is made when a solid or a liquid is dissolved in another liquid. This means that the solid mixes with a liquid so that you cannot see that there are two items. If a liquid is heated, the molecules in both items move faster, causing the solid to dissolve faster. If you separated a solution, you would see crystals.

You will need:
2 clear plastic cups
2 spoons
hot and cold water
sugar

1. Fill one cup with hot water.
   Fill the other with cold water.

2. Have a classmate help you now.
   Put a spoonful of sugar in each cup at the same time.

3. Stir both solutions at the same speed. Watch to see what happens.

Answer the questions.

1. What does *dissolve* mean? _____

2. In which cup did the sugar dissolve first? Explain.

   _____

   _____

3. Why is the sugar water a mixture? _____

   _____

4. How could you separate the sugar water solution?

   _____

Unit 1: What's the Matter?

Name _____ Date _____

Matter in Water and Air

A solution is formed when a solid or a liquid is dissolved in another liquid. In this activity, you will learn how different types of solids or liquids affect how things dissolve.

You will need:
4 bouillon cubes
5 clear plastic cups
hot and cold water

NOTE: This experiment must be done with an adult.

1. Place a bouillon cube in two of the cups. Then crumble the other two cubes and put them into two of the other cups.

2. Put cold water in a cup that has a whole cube in it and in one that has a crumbled cube in it.

3. Using hot water, repeat step 2 with two more cups.

4. Pour a small amount of dissolved bouillon solution into the last cup, and set it in a sunny place.

Answer the questions.

1. In which cup did the bouillon dissolve first? _____

2. How does crumbling a solid affect how it will dissolve in a liquid?

3. How does hot water affect a solid or a liquid that is dissolved in it? _____

4. What did you see in the bottom of the cup you put in a sunny place after the water had disappeared? _____

Unit 1: What's the Matter?

Separating Mixtures by Freezing

Mixing a drop of food coloring in a glass of water is easy to do. Separating the mixture of water and coloring is not so easy. Yet there is a way to do it.

You will need:

red or blue food coloring	paper cup	water
paper towels or a tray	measuring cup	freezer
spoon		

1. Pour 1/2 cup of water into the paper cup. Add one drop of food coloring. Stir. What happened to the water?

   _____

2. Put the cup in the freezer. Leave it overnight or until it is frozen solid.

3. Run some water over the outside of the paper cup. This will make it easy to remove the ice.

4. Put the ice cube on a tray or on a paper towel. Look at the ice cube.

GO ON TO THE NEXT PAGE

Unit 1: What's the Matter?

Name _____ Date _____

Separating Mixtures by Freezing, p. 2

Answer the questions.

1. Did the entire ice cube turn color? _____

2. What parts of the ice cube turned color? _____
   _____
   _____

3. What is in the clear part of the cube? _____

4. What is in the colored part? _____
   _____

How Did It Happen?
Why did the color separate from the water? Fresh water freezes at 0° Celsius. The coloring freezes at a slightly lower temperature. At first, it was cold enough only for the plain water to freeze. Some of the coloring separated out from the mixture. Some of the plain water froze. Then the temperature became lower. The rest of the mixture froze, too.

Unit 1: What's the Matter?

Name _____ Date _____

Separating Mixtures by Dissolving

Ink is a mixture of different substances.
You can separate the color of ink to show it.

✔ **You will need:**
felt-tip pen (purple or other dark color)   scissors
clear plastic cup   ruler
coffee filter paper (or paper towel)   pencil
markers or crayons

1. Cut a strip of coffee filter paper about 3 cm wide by 9 cm long. Put a drop of ink about 2 cm from one end. Let it dry.

2. Make a hole near the other end of the strip. Put a pencil *through* the hole.

3. Set the strip over the cup. Using the felt-tip pen, put a mark on the outside of the cup where the strip ends.

4. Remove the strip. Fill the cup with water to just above the mark. Put the strip back in the cup. The water should reach the bottom of the strip. It should not reach the ink spot.

5. Wait about 15 minutes.

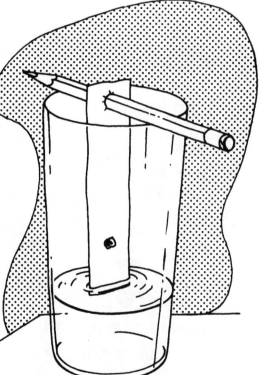

GO ON TO THE NEXT PAGE ➡

Unit 1: What's the Matter?

Separating Mixtures by Dissolving, p. 2

Answer the questions.

1. What happened to the water? _____

2. What happened to the ink? _____

3. What color moved up the strip first? _____

   Last? _____

4. Draw a picture to show what happened.

5. In a mixture, do the molecules of the substances change?
   _____

6. What other mixtures could you separate? _____
   _____

How Did It Happen?
Ink is a mixture of different substances. In water, these substances dissolve. They each dissolve at their own speed, however. The one that dissolves first moves up the strip ahead of all the others. The one that dissolves last stays near the bottom.

Unit 1: What's the Matter?

Name _____ Date _____

Matter in Air

Air is a mixture made of several gases. Oxygen is one of the gases in air. It takes fuel, oxygen, and heat to make a fire burn. The oxygen changes the fuel into a different kind of gas.

✔ **You will need:**

pie plate	candle	water	small glass jar
large glass jar	stopwatch	matches	pencil

NOTE: This experiment must be done with an adult.

1. Stand the candle in the pie plate. Fill the pie plate with water. Have the adult light the candle for you.

2. Turn the small jar upside down over the candle. Use the stopwatch to count the seconds until the candle goes out.

3. Fill in the table.

4. Have an adult light the candle again. Turn the large jar over the candle. Use the stopwatch to count the seconds until the candle goes out.

5. Fill in the table.

Time Candle Burns

Jar	Seconds Until Flame Goes Out
Small	
Large	

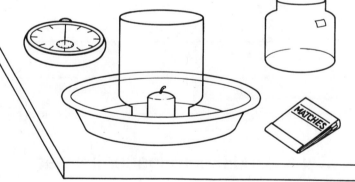

GO ON TO THE NEXT PAGE ➡

Unit 1: What's the Matter?

Name _____   Date _____

Matter in Air, p. 2

Answer the questions.

1. What does a fire need to burn?
   _____

2. What materials did you use that made the candle burn?
   _____
   _____

3. For how many seconds did the flame burn inside the small jar?
   _____

4. For how many seconds did the flame burn inside the large jar?
   _____

5. What was needed to make the candle burn?
   _____
   _____
   _____

6. Will a candle burn longer if it has more oxygen? Explain.
   _____
   _____
   _____

Name _____ Date _____

Matter in Water and Air

Oxygen is in the air. We need oxygen to breathe, and fire needs oxygen to burn. Also, oxygen makes iron and steel rust.

A. The pictures below show how oxygen is being used. Read the phrases written at the left. Then match each phrase with the picture it describes.

a.

1. To get energy to be able to grow, work, and play

_____

b.

2. To make heat and light

_____

c.

3. To change iron into rust

_____

B. If suddenly there were no oxygen, what would happen in each picture?

1. _____
2. _____
3. _____

Different Kinds of Change

A physical change is a change when the molecules of a substance do not change. The substance can be torn, cut, melted, evaporated, or condensed. A chemical change is a change when the properties of a substance change. The molecules become something else. For example, burning a log causes a chemical change. The log changes to ashes, smoke, and other kinds of gases.

You will need:
pie plate   rubber band   paper clip
scissors   matches   clock

NOTE: This experiment must be done with an adult.

1. Stretch the rubber band.

2. Cut the rubber band.

3. Make a "V" with the paper clip. Put the clip in the pie plate. Prop up the rubber band on the opened clip. Have an adult light the lifted part of the rubber band.

4. Let the rubber band cool for 15 minutes. Pick up the rubber band. Feel the burned part of the rubber band. Stretch the burned part of the rubber band.

Answer the questions.

1. What kind of change did you make when you stretched the rubber band? Explain. _____

2. What kind of change did you make when you cut the rubber band? Explain. _____

3. What kind of change did you make when the rubber band was burned? Explain. _____

Unit 1: What's the Matter?

Name _____ Date _____

What Kind of Change?

Each of these pictures shows a physical change or a chemical change.

Write *physical* or *chemical* under each picture to show the kind of change.

1.
_____

2.
_____

3.
_____

4.
_____

5.
_____

6.
_____

Name _____ Date _____

A Cooking Change

A physical change is a change when the molecules of a substance do not change. A chemical change is a change when the properties of a substance change. The molecules become something else in looks, taste, and feel.

✓ **You will need:**
box of oatmeal	margarine	eggs	brown sugar
white sugar	soda	vanilla	raisins
measuring spoons	large bowl	spatula	pot holders
measuring cup	mixing spoon	baking trays	

NOTE: This experiment must be done with an adult.

1. Follow the recipe on the back of the oatmeal box to make oatmeal cookies.

2. Add 1 cup of raisins to the cookie dough.

3. Drop dough by a tablespoon onto a baking sheet. Follow the recipe to bake the cookies.

Answer the questions.

1. What happened to the sugar, flour, and margarine when baked? Explain. _____
_____

2. What caused the change? _____

3. What happened to the raisins? Explain. _____
_____

4. If the cookies had burned, what kind of change would you have seen? Explain. _____
_____

Name _____ Date _____

A Model Fire Extinguisher

The oldest type of fire extinguisher contains a mixture of baking soda and water. It is called the soda-acid extinguisher. Near the top is a small bottle of acid. When the extinguisher is turned upside down, the acid mixes with the baking soda. A chemical change takes place. A gas forms. The gas pushes the water out through the hose. The water puts out the fire.

☑ **You will need:**
- baking soda
- water
- vinegar
- hammer
- tall narrow jar
- large jar with a screw-top lid
- nail
- measuring spoon
- measuring cup

NOTE: This experiment must be done with an adult.

1. Take the lid off the large jar. Have your teacher use the hammer and nail to punch three large holes close together on the lid.

2. Pour 2 cups of water into the large jar. Put the lid on the jar. Hold the jar upside down over the sink. What happens? _____
_____
_____

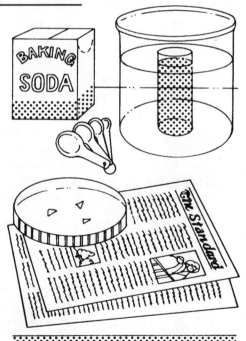

GO ON TO THE NEXT PAGE ➤

Unit 1: What's the Matter?

Name _____   Date _____

A Model Fire Extinguisher, p. 2

3. Empty the jar and refill it with 2 cups of water. Fill the narrow jar with vinegar and place it inside the large jar. The top of the narrow jar should be above the level of the water. Try not to spill any of the vinegar into the water.

4. Add 2 tablespoons of baking soda to the water in the large jar. Stir.

5. Securely screw the lid on the large jar.

6. Stand over the sink. Carefully turn the jar upside down. The vinegar should spill into the baking soda mixture. What happens? _____

Soda-acid fire extinguishers work on small paper and wood fires. They cannot be used on grease fires. The water will spread the grease and the fire. They cannot be used on electrical fires, either. Electric current travels through water. The current can cause a dangerous electric shock.

Answer the questions.

1. What kind of change happened between the baking soda and vinegar? Explain. _____

_____

2. If the vinegar and water had mixed in step 3, what kind of change would you have made? Explain. _____

_____

Name _____ Date _____

Unit 1 Science Fair Ideas

A science fair project can help you to understand the world around you. Choose a topic that interests you. Then use the scientific method to develop your project. Here's an example:

1. **Problem:** What happens to a steel-wool pad when it is left outside in the rain?

2. **Hypothesis:** The steel-wool pad will rust.

3. **Experimentation:** Materials: 2 steel-wool pads, water, 2 trays
   • Place a steel-wool pad in each tray.
   • Sprinkle water on one steel pad.
   • Place both trays uncovered on a shelf for two weeks.

4. **Observation:** The steel-wool pad begins to rust.
   The other pad has not changed.

5. **Conclusion:** Water make the steel in a steel-wool pad rust.

6. **Comparison:** The conclusion and the hypothesis agree.

7. **Presentation:** Research to find why the steel-wool pad rusted. Then prepare a presentation or a report to explain your results. Display both trays, showing the rusty pad and the unchanged pad.

8. **Resource:** Tell the books you used to find why the steel-wool rusted. Tell who helped you to get the materials and set up the experiment.

Other Project Ideas

1. What are the little bubbles in a can of soda? Do research to find out what the bubbles are.

2. What is the easiest way to open a jar lid when it is stuck? Set up and perform an experiment trying different ways to open the jar lid.

3. How could you show that lemonade is a solution? Do research and then experiment to show how to break down lemonade into its separate ingredients.

Unit 2: The Force is With You!

INTRODUCTION

Students in third grade are beginning to question how and why things work. They observe a variety of actions, reactions, and machines, and they can now understand the beginning concepts of force and motion. This unit will give information on force, gravity, friction, and inertia. It will also explain the six kinds of simple machines–levers, pulleys, inclined planes, screws, wedges, and wheels and axles.

FORCE

A force is simply a push or a pull. There are magnetic forces and gravitational forces, but in this unit, the forces associated with gravity will be explored. (Magnetic forces will be explored in Unit 3.) Forces can be balanced or unbalanced, and it is the interactions of these kind of forces that create motion. If forces are balanced, there is no movement. For example, if a kite is not moving while you are flying it, the force of the wind is balancing the weight of the kite and the pull of the string. However, if the wind stopped blowing, the weight of the kite and the pull of the string would be greater, causing the forces to become unbalanced. The kite would then move.

Forces also differ in size and direction. To move a book, it takes a small amount of force; but to move a bookshelf, it would take much more force. Forces can also come from up, down, left, and right. A force can be measured. Sometimes it is hard for students to understand that when pushing or pulling an object, even though the object does not move, the push or pull can be measured.

Force is measured in newtons. They can be added and subtracted. If forces are going in the same direction, they are added. For example, if someone is pushing a wagon and another person is pulling the wagon, the amount of forces being exerted can be added together. However, if people are pulling in opposite directions, say in a tug-of-war, the forces would be subtracted. The team having the greater number of newtons would have a greater force and would win.

Gravity

Gravity is a force that attracts all objects that have mass. It is the force that keeps all objects from flying off the surface of the Earth. It is also the force that keeps the planets, Moon, and stars in orbit. Everything on Earth is pulled to the center of the Earth by this unseen force. Sir Isaac Newton called this force gravitation; and he explained that all things have force, but that the pull of the Earth is greater, thus anchoring objects to the Earth's surface and making things fall down. The more massive an object, the greater the force that will be exerted. The force of gravitation is about 9.8 newtons per kilometer for every object on Earth.

Gravitational force depends on the mass of an object and how far apart the centers of the objects are. The more mass an object has, the greater the gravitational pull will be. The gravitational force between the Earth and other objects is greater because the mass of the Earth is so large. If the mass of two objects is small, the gravitational force will also be small because the force of the Earth's gravity is greater. For example, if two books are side by side on a table, they will stay on the table because of the pull of gravity from Earth. If the forces of gravity and friction were not working, the books would move toward each other. Furthermore, the farther apart the bodies are, the less the pull will be because less force can be exerted.

Another concept difficult for students to understand is the difference in the terms *mass* and *weight*. *Mass* is the measure of the amount of matter in an object. As discussed in Unit 1, mass is measured in grams (g). *Weight* is the measure of the force of gravity on an object. A spring scale measures weight using newtons. When students step on a scale, they are actually measuring their mass, since weight is measured in newtons. It can best be explained by comparing the mass and weight of a person on Earth and on the Moon. The mass of the person stays the same in either place. However, the weight of the person on the Moon will be one sixth of the weight on Earth. The gravitational pull is one sixth less on the Moon since there is less mass with the Moon.

MOTION

The motion of an object is the result when a variety of forces interact. A change in motion occurs if a still object moves, or an object already in motion changes speed or direction. Two different forces, acting in opposite directions, will interact so an object will not move. These forces are considered balanced forces. An unbalanced force results when a force is placed on an object either at rest or in motion, making the object change its state. The object will move faster as the forces become more unbalanced.

Suppose a soccer ball is on the field. It is in a state of rest; the forces are balanced. But if someone kicks the ball, the force of the kick makes the ball move. The greater the force, the faster the ball will go, and the farther it will go. But there is more to motion than just force. The movement of an object is also affected by friction, the force that resists movement.

Friction

Friction is a force that keeps resting objects from moving and tends to slow motion when one object rubs against another object. Every motion is affected by friction. It is useful when movement needs to be slowed, but it causes problems when something needs to be moved. An object's surface determines the amount of friction. Rough surfaces, like concrete, dirt, and grass, create more friction. It is helpful when walking, so that people can walk without sliding. These surfaces create a problem when trying to move something over them. Smooth surfaces, like ice and lacquered wood, have less friction, so motion would be easier. Moreover, heat is produced as the objects rub across each other. Early humans used the force of friction to make fire when they rubbed two sticks together.

Mass and surface areas of objects affect the amount of friction. The heavier an object is, the greater the amount of friction. By lightening the load, friction will be decreased, and the load can be moved easily with less friction. Similarly, when large surface areas come into contact during motion, friction is greater. By reducing the contact of the surface areas, the object can be moved more easily.

In some cases, friction can be reduced by using lubricants, materials like oil or soap. Lubricants coat the surface of an object to decrease rubbing. Machines need lubricants to reduce the friction when parts rub against each other. This helps to keep the parts cool to avoid fire hazards as well as to keep them moving smoothly.

Inertia

Inertia is the tendency for all objects to stay still or to keep moving if they are moving. There is only a change in the speed or direction when an outside force acts on the object. This concept is known as the *first law of motion,* or *the first law of inertia,* and was explained by Sir Issac Newton in the 1700s.

Inertia is a property of mass. The more mass an object has, the more it resists a change in its state. Likewise, the greater the mass, the greater the inertia. Once an object is moving, it tends to maintain its direction and speed. It will continue to move in a straight line unless acted upon by another, unbalanced force. A person riding in a car travels at the same speed as the car. The person has the same rate of inertia. But if the car stops, the person's inertia tendency is continued at the same speed and direction. The seat belt acts as an outside force to stop the forward movement.

MACHINES

Machines are devices that help make work easier. They usually change the amount or direction of forces needed to get a job done by lifting, pulling, pushing, or carrying. Friction usually affects the efficiency of a machine, because the amount of effort is decreased. Some machines are able to decrease the friction, and thus increase the ease of work. There are six kinds of simple machines: levers, pulleys, inclined planes, wedges, screws, and wheels and axles. When one or more of these are joined in one machine, it is a compound machine. The end result is that less force is needed to do work.

Simple Machines

A simple machine has no moving parts or only a few moving parts. Simple machines are made with materials close at hand.

Levers

A lever is a bar or board that rests on a point, called a *fulcrum*. If you push down on one end of the bar, the other end rises. The direction of the force is changed. By using a lever, heavy objects can be lifted because less force is needed.

Most levers place the fulcrum in the center. The load and force are at each end. This is known as a first-class lever. When using levers to lift loads, the force is always applied down. Moreover, the closer the fulcrum is to the load, the easier is the work. A seesaw is an example of a first-class lever. A second-class lever places the fulcrum on one end, the load in the middle, and the force at the other end. The force in this kind of lever pulls up. A nutcracker is a second-class lever. A third-class lever has the fulcrum and load on each end, and the force is applied in the middle. A pair of tweezers is an example of this kind of lever.

Most commonly recognized levers are first-class levers. Ones that third graders will recognize are bottle openers, pliers, crowbars, and spatulas.

Pulleys

A pulley is a wheel with a rope wrapped around it. The wheel generally has a groove in it to keep the rope from slipping off. It is also a kind of lever used to lift heavy things. One end of the rope is tied to the object and the other end is pulled by a person or machine. The wheel turns freely so there is little friction with the rope. Using several pulleys in one machine uses less effort to lift something.

There are two kinds of pulleys, fixed and movable. A fixed pulley stays in place as the load is lifted. When someone pulls on the end of the rope, the direction of the force is changed. The work is easier because the direction of the force is changed. Flag poles, window blinds, and sailboats use fixed pulleys. In movable pulleys, the pulley is attached to the load, leaving it free to move with the load. A movable pulley is lifted. It makes work easier because less force is needed. Cranes often use movable pulleys.

Inclined Planes

An inclined plane is a flat surface that slopes. Even though the distance to move an object is greater, less work is required to move an object up an inclined plane than to lift it. A ramp is the most common kind of inclined plane. The angle of the plane also affects the force needed to move an object up a ramp. The steeper the plane, the more force is needed to move an object.

Friction is a force that can affect the ease of using an inclined plane. Flat objects being pushed or pulled up an inclined plane would be harder to move up the slope. More force would be needed to move them. Decreasing the surface area of the object or changing the surface area of a ramp would make the work easier.

Wedges

A wedge is a kind of inclined plane. It is made when two inclined planes are joined together to form one sharp edge. Wedges are often used to break something into two parts. The force is applied at the point, giving a greater force to make the work easier. A thinner wedge would not need as much force as one that is larger. Axes, forks, knives, and needles are examples of wedges.

Screws

A screw is an inclined plane that wraps around a rod to make a spiral. The edge or ridge of the screw is known as the thread. The thread moves between the wood to break it apart as the screw is turned. Like other inclined planes, screws decrease the amount of force needed to work, but increase the distance needed to move. Screws often are used to fasten things together, make holes, or lift objects. Other examples of screws are the stem of a car jack, power drills, pencil sharpeners, and spiral staircases.

Wheels and Axles

A wheel and axle is another kind of lever. It uses a handle to turn around the rod, a fulcrum. When the wheel turns, the rod turns; and when the rod turns, the wheel turns. The axle usually goes through the center of the rod. By turning the axle, speed and distance are increased, but more force is needed. By turning the wheel, more force is gained, but the speed and distance are decreased. Moreover, the larger the wheel, the more it will need to be turned to complete the job. In this case, less force is needed. Examples of axles and wheels include doorknobs, screwdrivers, fishing reels, and pencil sharpeners.

Compound Machines

Compound machines are made up of two or more simple machines. Most machines are compound machines. By using a combination of simple machines, one can apply more force to make the work even easier. The force is increased as it passes from one machine to another. A bike is an example of a compound machine. Wheels and axles, screws, and levers work together to make the bike work.

Friction is a factor in a machine's movement, because this force works against the motion of the machine. The machine slows down as the parts rub against each other. In a bike, the rubber of the brakes rubs on the rim of the metal wheel to slow down the bike.

Name _____   Date _____

Push or Pull?

A force is a push or pull. Where there is force, there is motion.

Write *push* or *pull* for each picture.

1. _____

2. _____

3. _____

4. _____

5. _____

6. _____

© Steck-Vaughn Company   64   Unit 2: The Force Is With You!
Physical Science 3, SV 3762-3

Measuring Forces

A force is a push or pull. Forces can be measured. Just as distances are measured in units called meters, forces are measured in units called newtons. When you push or pull on something, you apply a force to it.

Match each item with the unit that is used to measure it.

_____ 1. distance   **a.** newtons

_____ 2. mass   **b.** meters

_____ 3. time   **c.** seconds

_____ 4. force   **d.** grams

5. Mark and Suzanne each have a spring scale. They have hooked the scales together. They are both pulling on their spring scales, but the spring scales are not moving. The reading on Mark's spring scale is 4 newtons. What is the reading on Suzanne's spring scale? Explain your answer. _____

_____

Look at the spring scales shown below and answer the following questions.

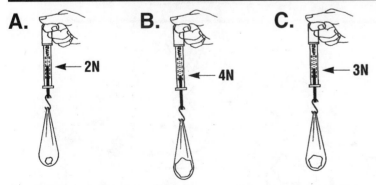

6. Which of the spring scales is holding the heaviest load? _____

7. Which spring scale is holding the lightest load? _____

Unit 2: The Force Is With You!

Name _____ Date _____

More and More Force

The more something weighs, the more force will be needed to move it.

Color the picture that takes the most force to move.

Unit 2: The Force Is With You!

Name _____   Date _____

Calculating Force

Forces can make things move and can change the motion of things that are moving. Two or more forces can act at the same time. If the forces act in the same direction, they have a greater effect. If they act against each other, they have a lesser effect. Forces can be added and subtracted using the scientific measurement of newtons. Use what you know about forces, addition, and subtraction to answer these questions.

Answer the questions.

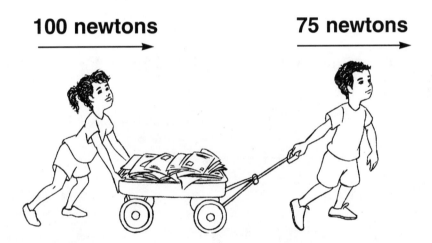

1. Sarah and David are trying to move a wagon. David is in front of the wagon and is pulling with a force of 75 newtons. Sarah is behind the wagon and is pushing with the force of 100 newtons. What is the total force acting on the wagon? Which way will the wagon move? _____

2. Suppose Sarah starts pulling on the back of the wagon with a force of 100 newtons instead of pushing it. Which way will the wagon move? Why? _____

GO ON TO THE NEXT PAGE

Unit 2: The Force Is With You!

Name _____   Date _____

Calculating Force, p. 2

3. Some students are having a tug-of-war. Chris's team is pulling with a force of 250 newtons. Emily's team is pulling with a force of 260 newtons. Which way will the rope move? _____

4. Emily is pulling with a force of 60 newtons. But she slips and lets go of the rope. How much force is her team now pulling with? Which way will the rope move? Explain your answer. _____

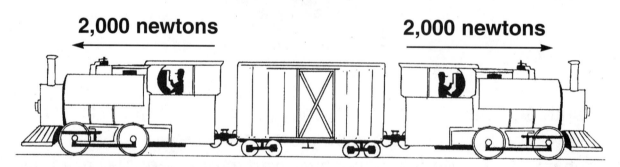

5. A locomotive is hooked onto each end of the boxcar. Each locomotive is pulling on the car with a force of 2,000 newtons. Will the boxcar move? Explain your answer. _____

6. Suppose the locomotive on the left is unhooked from the boxcar. What is the total force on the boxcar now? Which way will the boxcar move? _____

Gravity

Another force that acts between objects even when they are not touching is gravity. All objects pull on other objects. Earth's gravity pulls on everything on Earth's surface and in the atmosphere. It also pulls on everything else in the solar system and the universe. The force of Earth's pull on an object can be measured. This measurement is the object's weight. You have weight because Earth's gravity is pulling on you. If you were on another planet, your weight would be different. However, your mass would be the same, because the mass of an object is the measure of the amount of matter in the object.

Circle the best answer.

1. Magnetism is a force that can act between objects that aren't touching. Another force that works the same way is . . .

    a. wind. **b.** gravity. **c.** friction. **d.** inertia.

2. Gravity is a . . .

    a. lifting force. **b.** squeezing force. **c.** pushing force. **d.** pulling force.

3. The measurement of the pull of Earth's gravity on an object is its . . .

    a. length. **b.** mass. **c.** weight. **d.** magnetism.

4. The mass of the Moon is about $\frac{1}{6}$ the mass of Earth. Therefore, the force of gravity on the Moon is $\frac{1}{6}$ the force of gravity on Earth. Use this information and what you know about weight and mass to complete the chart below. The first one has been done for you.

Mass on Earth	Weight on Earth	Mass on Moon	Weight on Moon
3 kilograms	30 newtons	3 kilograms	5 newtons
6 kilograms	60 newtons		
9 kilograms	90 newtons		
12 kilograms	120 newtons		
15 kilograms	150 newtons		

Unit 2: The Force Is With You!

Name _____ Date _____

Is Gravity Working?

Gravity is the force that pulls objects to the Earth.

Color the things that are wrong in the picture.

Unit 2: The Force Is With You!

Name _____ Date _____

Uphill and Downhill

Moving uphill is harder since an object is trying to resist the pull of gravity. Going downhill is easier since gravity helps pull an object toward Earth.

Color each picture that shows objects trying to resist gravity.

1.

2.

Name _____ Date _____

Which Takes More Force?

It takes force to lift something, because gravity pulls everything toward Earth. An object that has more mass will need more force to be lifted.

Color the picture in each set that takes more force to lift.

1.

2.

3.

4.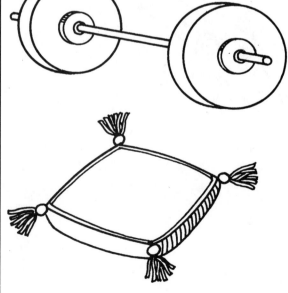

Unit 2: The Force Is With You!

Name _____ Date _____

The Force of Gravity

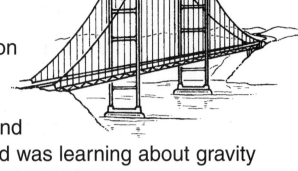

Ted walked across a bridge every day on his way home from school. The bridge crossed a river, and sometimes he would drop a pebble or a small stone and watch it splash into the water below. Ted was learning about gravity in school, and on his way home one day, he began to wonder if heavier objects fall faster than lighter objects. He thought they did, but he wasn't sure. He decided to do an experiment.

When he reached the bridge, he collected several stones and lined them up from lightest to heaviest. He then dropped the lightest stone off the bridge and timed how long it took to fall to the water. By using a stopwatch, he found that it took 5 seconds for the stone to hit the water.

Ted then chose a stone that was about twice as heavy as the first one. "I predict this stone will fall faster," he thought. He dropped it over the edge and timed it. Ted was surprised to find out that it took 5 seconds to fall, the same amount of time the first stone had taken.

"Maybe I didn't use a heavy enough stone," he thought. He grabbed one of the heaviest stones. "This one will not take as long as the other two did," he predicted. He dropped it over the edge and timed how long it fell. It also took 5 seconds to hit the water.

Ted sat down and thought for a minute. "Maybe I was wrong," he said to himself. After thinking things over, he made a new prediction. He found a very heavy rock. It was heavier than any other he had tried. He dropped it off the bridge and timed how long it fell. When it hit the water, he smiled and said, "I got it right!" Satisfied with his experiment he headed home.

GO ON TO THE NEXT PAGE

Unit 2: The Force Is With You!

The Force of Gravity, p. 2

1. What was Ted trying to prove?

_____

_____

2. What experiment did he perform?

_____

_____

3. How long did it take for the last rock to hit the water?

_____

4. After observing and thinking for a while, Ted revised his prediction. What new information did he have that would make him want to change his prediction?

_____

_____

5. What did Ted learn about gravity?

_____

_____

Name _____   Date _____

What Can More Mass Do?

The more mass an object has, the more force will be needed to lift or move it.

 You will need:
- paper clip
- paper cup
- 8 washers
- 2 pieces of string (each about 80 cm long)
- transparent tape
- measuring tape
- stopwatch

1. Unbend the paper clip, and tape it to the cup.

2. Tie two washers to the end of one string. Tie a loop at the other end of the string.

3. Place the cup on a table, 60 cm away from the edge. Mark the spot with a piece of tape. Put one washer in the cup. While your partner holds the cup, hook a string onto the paper clip. The string should lie across the table, and the washers should hang over the opposite edge.

4. Get the stopwatch ready. When your partner lets go of the cup, time how long it takes for the cup to slide to the edge of the table. Record the time in the chart on the next page.

5. Put the cup back where it was, and put another washer in it. Repeat step 4. Do this until all the washers are in the cup.

GO ON TO THE NEXT PAGE ➤

Unit 2: The Force Is With You!

Name _____ Date _____

What Can More Mass Do?, p. 2

TIME THE CUP TOOK TO SLIDE ACROSS THE TABLE

Number of Washers in the Cup	Time
1	
2	
3	
4	
5	
6	

Answer the questions.

1. Look at the information you wrote in the chart. What happened when you added more washers?

   _____
   _____
   _____

2. What do you think would happen if you kept adding more and more washers to the cup before you let it go?

   _____
   _____
   _____

3. Based on what you found out in this experiment, what would you do if you wanted to keep a basket from blowing away in a windstorm?

   _____
   _____
   _____

Unit 2: The Force Is With You!

Name _____ Date _____

What Causes Motion?

Suppose you tied a string to a book and hung the book from a spring scale. Earth's gravity would pull down the book. The spring in the spring scale would pull up on it. The two forces acting on the book would be equal and acting in opposite directions. Since the forces would be balanced, the book would not move. If you cut the string, the spring would no longer pull on the book, but gravity still would. The forces would no longer be balanced, so the book would fall to the floor. Whenever an unbalanced force acts on an object, it moves or its motion changes.

Answer the questions.

1. Teams A and B are having a tug-of-war. The rope is not moving. Therefore, Team A is pulling _____ Team B. Circle the phrase that completes the sentence.

   a. with more force than

   b. with less force than

   c. with the same force as

2. Suppose you had a heavy box that you wanted to move to the other side of the room. You tried pushing it by yourself, but you couldn't move it. Explain how having a friend also push might help. _____

   _____

   _____

   _____

Unit 2: The Force Is With You!

Name _____ Date _____

Where Does It Go?

Inertia is the tendency for all objects to stay still if they are still or to keep moving if they are moving. The motion will only change if an outside force acts on the object.

 You will need:
marble   shoe box

1. Place the box on a table. Put the marble in the box so that it touches the right wall of the box.

2. What will happen if you slide the box to the right? Move the box gently to the right. (If you move the box too quickly, the marble will bounce.) Was your prediction correct?

3. Now, where will the marble move if you slide the box to the left? Move the box gently to the left. Was your prediction correct?

4. Put the marble in the middle of the box. How will the marble move if you move the box away from you? Try it.

5. Try placing the marble in other places in the box. Predict where the marble will go.

 Answer the questions.

1. Did the marble move when you moved the box? Explain.

   _____

   _____

2. What property did you see? _____

Unit 2: The Force Is With You!

Friction

If you rolled a ball through the grass, the ball would rub against the blades of grass it touched. The force of this rubbing would slow the ball down until it came to a stop. This "rubbing" force is called friction. If you then rolled the ball across a parking lot, there would be less material rubbing against the ball. Therefore, the ball would roll faster and longer. Still, it eventually would come to a stop.

Read each list of surfaces. Suppose a small ball rolled across each surface.

Circle the surface in each row that would best slow down the ball.

1. sandpaper   wax paper   carpet   grass
2. thick carpet   thin carpet   tile floor   wood floor
3. tile floor   rubber mat   dry bathtub   towel
4. sheet of ice   cement   gravel   brick

Circle the best answer.

5. A person adds oil to a car's engine to . . .

   a. reduce friction.   **b.** increase friction.
   c. reduce inertia.   **d.** increase inertia.

6. Suppose you set a heavy box on a sheet of wax paper. How might that make it easier to move the box?

   _____
   _____
   _____
   _____

Name _____   Date _____

Friction and Lubricants

Have you ever rubbed your hands together on a cold day? What happens? Try it now. When things rub together, they get warm. This is called friction.

Many machines have parts that rub against each other. If they rub too much, they might become too hot. This could cause the machine to break down or start a fire. How can friction be reduced? A lubricant can be used. Oil is most often used as a lubricant. It makes a thin film on each piece of the machine. This reduces the amount of rubbing. Try the following activity.

✓ **You will need:**
a jar with a cover   soap   water

1. Screw the cover as tightly as possible on the jar. Unscrew it. Then tighten it again.

2. Using soap and water, wet and lather your hands. Try to unscrew the cap. Why is it more difficult?

   _____

   _____

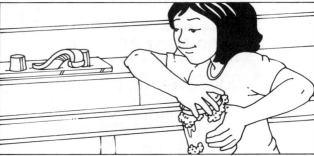

 Answer the questions.

1. Which way were you able to unscrew the jar lid easily? Explain.

   _____

   _____

2. What lubricant did you use when trying to unscrew the lid in step 2? _____

Unit 2: The Force Is With You!

Name _____ Date _____

Testing Friction

Adam's baby sister Kate was learning to walk. Sometimes, she would slip and fall down. Adam noticed that Kate fell a lot when she wore plastic sandals. She fell less often when she wore sneakers. He also noticed that she fell more often on the wood floor in the hall than on the carpet in his bedroom.

Adam thought that Kate's sneakers gripped the floor better than her sandals did. He also thought the wood floor was more slippery than the carpet was. Adam thought he knew why Kate fell more often under certain conditions. But he wanted to be sure. He decided to do an experiment. He gathered Kate's sneakers, her sandals, a wooden floor tile, and a piece of carpet. He propped up one end of a large book to make it slant.

For each trial, Adam covered the book with either the wood, the carpet, or both materials next to each other. Adam allowed two shoes to slide down the material, one at a time. His results are shown in the table below.

Sliding Sneakers and Sandals

Trial	Object	Surface	Which slid faster?
1	Sandal	Wood	X
	Sneaker	Wood	
2	Sandal	Carpet	X
	Sneaker	Carpet	
3	Sandal	Carpet	
	Sandal	Wood	X
4	Sneaker	Carpet	
	Sneaker	Wood	X

GO ON TO THE NEXT PAGE ▶

Unit 2: The Force Is With You!

Testing Friction, p. 2

1. What was Adam's hypothesis?

_____

_____

_____

2. Which shoe made more friction? How do you know?

_____

_____

_____

3. Which surface made more friction? How do you know?

_____

_____

_____

4. How did Adam's experiment help him to figure out why his sister was falling?

_____

_____

_____

Unit 2: The Force Is With You!

Name _____ Date _____

Which Is the Easy One?

Friction is the force that acts between two objects that touch or rub against each other. It is a force that slows or stops movement.

Color the picture in each set that shows which object is easier to move. Tell why you chose the picture you did.

1.

_____

2.

Unit 2: The Force Is With You!

Reaction Actions

Suppose a ball was flying through the air and there was no such thing as friction. What if no forces pulled or pushed the ball? It would just go on moving forever and ever. The ball, or any other object, would maintain its motion until some outside force acted on it. This is a property that all objects—and all matter—share.

This property is called inertia. You can observe inertia in action when you ride in a car. When the car starts moving, you feel the back of the seat push against you so that you start moving, too. If the car stops suddenly, you will feel the seat belt pull against you and slow you down.

Circle the best answer.

1. An object that is moving continues to move unless a force acts on it, because of . . .

   a. friction.   **b.** inertia.   **c.** gravity.   **d.** magnetism.

2. Suppose you were in a parked bus and there was a tennis ball on the floor of the bus. What would happen to the ball if the bus started moving forward? Explain your answer.

   _____

   _____

3. Suppose the tennis ball were on the floor of the bus while the bus was going down the road. What would happen to the ball if the bus driver suddenly stepped on the brakes? Explain your answer. _____

   _____

Unit 2: The Force Is With You!

The Law of Motion

The *first law of motion*, also known as the *law of inertia*, states that an object will continue to move or to remain at rest unless it receives an unbalanced push or pull.

The pictures and captions below show what would happen to a person in a car crash if he or she didn't wear a seat belt, but the pictures are out of order. Number the pictures in the correct order.

_____ 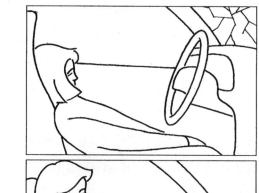 She bounces off the steering wheel and the windshield and lands back in her seat.

_____  What happens when a person is in a stopped car and the driver presses down on the gas pedal? The seat pushes her forward with the car.

_____  She flies into the steering wheel and the windshield.

_____  If the car stops suddenly in a crash, she keeps moving forward.

Unit 2: The Force Is With You!

Name _____ Date _____

Machines

Suppose you wanted to lift a very large rock out of a hole in the ground. Could you lift it out with your bare hands? Probably not. But you could get it out if you used a lever. One example of a lever is a bar resting on a point called a fulcrum. You could lift a rock out of a hole by placing one end of the bar under the rock and pushing down on the other end.

There are other types of levers as well. A wheel and axle can change the strength of a turning force. When you turn the wheel with the handle, the axle also turns. A pulley is a lever made up of a rope, string, or cable looped over a roller. When you pull down on one end of the rope, the other end moves upward.

Levers are simple machines. A machine is a tool that can change either the amount of force acting on an object or the direction of the force. A simple machine is a machine that has no moving parts or only a few moving parts.

Circle the simple machines.

1.

2.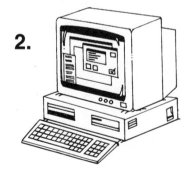

3.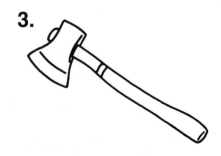

4.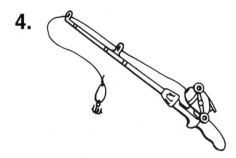

Unit 2: The Force Is With You!

Types of Levers

When you see two people on a seesaw, you are watching a first-class lever at work. The fulcrum is in the middle. The person who is closer to the ground is providing the effort. The person in the air is the load that is being lifted.

There are other classes of levers, too. Look at this diagram.

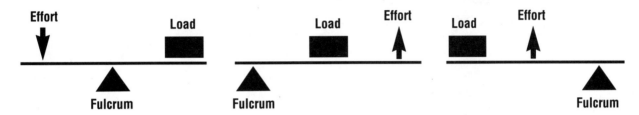

A second-class lever is one in which the load is in the middle, while the effort and fulcrum are at opposite ends. An example of this kind of lever is a bottle opener.

A third-class lever is one in which the effort is in the middle, with the load and the fulcrum at opposite ends. The arm of a person using a tennis racket to hit a ball is an example of a third-class lever. The shoulder is the fulcrum, the arm applies the effort, and the load, or resistance, is the ball hitting the racket.

Answer the questions.

Look at these drawings. Tell which class of lever each is.

1. Pliers

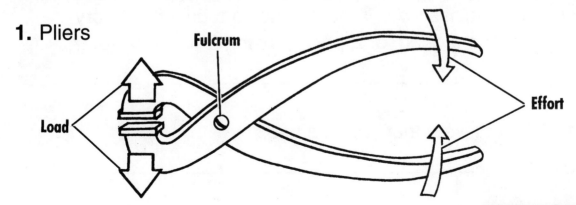

Name _____ Date _____

Types of Levers, p. 2

2. Nutcracker

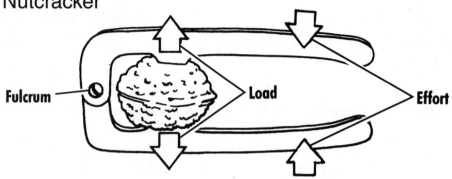

3. Tweezers

4. Look at the picture of the nutcracker. What will happen if the load force is stronger than the effort force?

   _____

5. Which is a first-class lever, a person using a crowbar to pry a rock out of the ground, or a person using a wheelbarrow?

   _____

6. Name 4 other levers that you can use.

   _____

   _____

A Lifting Machine

A lever is a simple machine made with a board or bar that moves on a fulcrum.

 You will need:
2 chalkboard erasers
sturdy meter stick
book

1. Place the erasers on top of each other. They will be the fulcrum. Place the meter stick on the erasers. The erasers should be 20 centimeters from one end of the stick. The meter stick will be the lever.

2. Place the book on the opposite end of the stick.

3. Push down on the end of the stick that is opposite the book. (Push gently so the book will stay on the stick.) Is it easy or hard to lift the book?

4. Move the fulcrum 10 centimeters closer to the book. Repeat steps 2 and 3, making sure the book is in the same place.

5. Move the fulcrum 10 centimeters closer to the book. Repeat steps 2 and 3. Keep moving the fulcrum to the book and trying to lift the book.

Answer the questions.

1. Which object is the load? _____

2. What happens when the fulcrum is moved closer to the load?

_____

_____

Unit 2: The Force Is With You!

Name _____ Date _____

Pulleys

A pulley is a simple machine made by wrapping a rope around a wheel. One end of the rope is tied to an object. By pulling on the other end of the rope, the object is lifted. Several pulleys can be used with one rope to lift a really big object.

 You will need:
2 broom handles
6 meter-long rope

1. Work in groups of three. Tie one end of the rope near the end of one broom handle.

2. Have two partners each hold a broom handle. Make sure they use both hands. Have them face each other and wrap the rope around both handles several times as shown.

3. Have your partners try to hold the handles apart as you slowly pull on the free end of the rope.

4. Repeat two more times so that each person gets to pull the rope.

 Answer the questions.

1. How are the broom handles and rope like a pulley?

   _____

   _____

2. How is the pulley changing the force?

   _____

   _____

Unit 2: The Force Is With You!

Name _____ Date _____

Inclined Planes

An inclined plane is an example of another simple machine. It is a flat surface that is raised at one end. It is often easier to push something up a ramp than it is to lift the same distance, because less force is needed.

✓ **You will need:**
- wooden board
- masking tape
- thin spiral notebook
- 30 centimeter-long string
- chair
- spring scale
- meter stick

1. Place one end of the board on the edge of the chair and the other on the floor. Tape the board to the floor to hold it in place.

2. Tie one end of the string around the notebook. Tie the other end to the hook on the spring scale.

3. Place the notebook on the floor and slowly lift it straight off the ground. Read the setting on the spring scale when the notebook is no longer touching the floor. Slowly, continue to lift the book until it is even with the seat of the chair. Record the number of newtons on the table on the next page.

4. Measure the height you raised the notebook. Record the distance.

5. Place the notebook on the floor at the base of the ramp. Slowly and steadily pull the spring scale up the ramp. When the notebook begins moving, read the scale. Continue to pull the notebook until it reaches the top of the ramp. Record the number of newtons needed to pull it up the ramp.

6. Measure the length of the ramp and record the distance on the table.

GO ON TO THE NEXT PAGE ➤

Unit 2: The Force Is With You!

Inclined Planes, p. 2

Force Needed to Move Object

	Without Ramp	With Ramp
Force		
Distance		

Answer the questions.

1. Which way of moving the book took more force?
   _____
   _____

2. Which way of moving the book took a greater distance?
   _____
   _____

3. How did the ramp help to move the book?
   _____
   _____

4. How might friction affect using a ramp?
   _____
   _____

5. Name three examples of uses of inclined planes.
   _____
   _____

Unit 2: The Force Is With You!

Wedges

A wedge is a simple machine formed when two inclined planes meet to make a sharp edge. Wedges are used to cut or break things apart. The force is placed on the point where the sharp edges meet, causing the object to be cut into two parts.

 Circle the pictures that are wedges.

1.

2.

3.

4.

5.

6.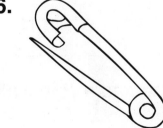

Unit 2: The Force Is With You!

Name _____ Date _____

Screws

A screw is another kind of inclined plane. It forms a spiral by winding around an inclined plane. Screws can hold things together, make holes, or help lift things. The screw forces the wood around the inclined plane to move apart, making a hole.

You will need:
8" x 6" sheet of construction paper
scissors   ruler   pencil

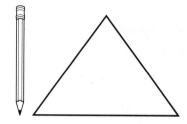

1. Using the ruler, draw a diagonal line from one corner of the paper to the opposite corner. Make sure you draw a very dark line.

2. Cut along the line to form two triangles.

3. Starting with the shortest edge of a triangle, wrap the paper around the pencil.

Answer the questions.

1. Where is the inclined plane of the screw?

   _____

   _____

2. How is the screw also a wedge?

   _____

   _____

3. How are a ramp and a screw alike? different?

   _____

   _____

   _____

Wheels and Axles

You can use a fork or an eggbeater to mix or beat something. Which one works better? With a fork, the force of your moving hand and arm equals the force of the moving fork. An eggbeater, however, uses a wheel and axle. The wheel and axle is a kind of lever that moves around a fulcrum. In a wheel and axle, the parts move together.

✓ **You will need:**
watch or clock with a second hand
fork   2 bowls   tape
eggbeater   soap flakes   water
measuring spoon

1. Fill both bowls about half full with water. Add one tablespoon of soap flakes to each bowl.

2. Have a classmate watch the clock. Beat one bowl of soapy water with a fork for about 15 seconds. Examine the suds.

3. Then beat the mixture in the other bowl for 15 seconds with the eggbeater. Examine the suds.

GO ON TO THE NEXT PAGE ➤

Unit 2: The Force Is With You!

Wheels and Axles, p. 2

4. Now find out how much an eggbeater multiplies force. Put a strip of tape on one blade as a marker.

5. Hold the beater so that the tape faces you. Turn the handle around once very slowly. How many times does the taped blade move past you? _____

_____

 Answer the questions.

1. How is the eggbeater like a lever? _____

_____

2. How many times did the blade move past you when you turned the handle? _____

3. In which bowl were the soap suds more frothy? Explain.

_____

_____

4. Name other machines that use wheels and axles.

_____

_____

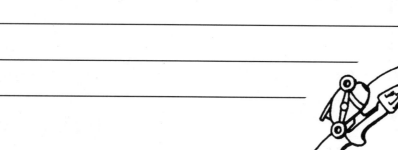

Unit 2: The Force Is With You!

Machines That Make a Difference

Jay broke his leg in a car accident. Dr. Crane, his doctor, told him, "The broken bones in your leg will heal faster if you don't move your leg around too much. So we will attach a rope to your leg and loop the rope over a pulley. A weight will be attached to the other end of the rope. The pulley will turn the downward pull of the weight into an upward pull on your leg."

A few weeks later, Jay's leg had started to heal. Dr. Crane told Jay he didn't have to stay in bed all the time. She gave him a wheelchair to use. After she helped him into the chair, he began to check it out. Unlike his parents' car or a shopping cart, which has four wheels of equal size, the wheelchair had two small wheels in front and two large ones in back. Jay asked the doctor why the wheels weren't all the same size.

"The rear wheels are larger on a wheelchair to make it easier for you to use," she told him. "To travel around in a wheelchair, you turn the handle–a ring–attached to each rear wheel. The bigger the handle is, the easier it is for you to turn the axle. But the ring can't be bigger than the wheel itself or it would hit the ground. Therefore, the rear wheel is made larger so the handle can be larger."

Jay started using his wheelchair to go places in the hospital. That was much better than staying in his room all day. After a while, he went down a hallway but had to stop. There were three stairs that went up to another hallway. "How am I supposed to get up those in a wheelchair?" he wondered.

GO ON TO THE NEXT PAGE

Unit 2: The Force Is With You!

Name _____ Date _____

Machines That Make a Difference, p. 2

A nurse walked by and noticed that Jay was looking disappointed. "What's the matter?" she asked.

"I can't get up the stairs in this wheelchair," Jay explained.

"Oh, that's no problem," she said. "Around the corner, there's a wheelchair ramp that you can use to get up there. A wheelchair ramp is an inclined plane. You have to go a little farther when you use it, but using it is a lot easier than trying to pull yourself and your wheelchair up the stairs."

Jay thanked her and went up the ramp. He enjoyed his trip around the hospital so much that he did it every day until he went home. After a few months, his leg finished healing and he was up and about on his own two feet.

During his hospital stay, Jay used the three simple machines listed below. Under each machine, describe how he used it.

1. Pulley

_____

_____

2. Wheel and axle

_____

_____

3. Inclined plane

_____

_____

Unit 2: The Force Is With You!

Compound Machines

A compound machine is a machine that is made up of two or more simple machines. A compound machine can increase the force to make work much easier.

 Answer the questions.

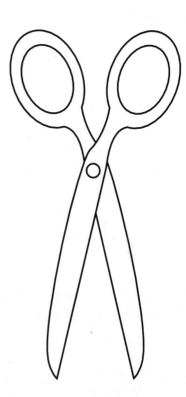

1. Label the simple machines that you see in the scissors.

2. Where is the fulcrum for the lever?
   _____

3. How does a pair of scissors make work easier?
   _____
   _____

4. Name other compound machines. Tell the simple machines that are in them. _____
   _____
   _____
   _____

Unit 2: The Force Is With You!

Name _____ Date _____

Unit 2 Science Fair Ideas

A science fair project can help you to understand the world around you. Choose a topic that interests you. Then use the scientific method to develop your project. Here's an example:

1. **Problem:** How does the size of a wheel and axle affect doing work?

2. **Hypothesis:** It is easier to turn a screw with a screwdriver that has a larger handle.

3. **Experimentation:** Materials: 2 screwdrivers of the same length, but handles of different sizes; 2 screws that are 3 centimeters long; wood block; hammer; tape
   - Use the hammer to set the screws in the wood block.
   - Put a thin piece of tape on the handles.
   - Use the smaller-handled screwdriver to turn one screw into the block. Count the number of turns it takes.
   - Use the larger-handled screwdriver to turn the other screw into the block. Count the number of turns it takes.

4. **Observation:** The screwdriver with the larger handle turned around fewer times to set the screw. It was also easier to turn than the screwdriver with the smaller handle.

5. **Conclusion:** In a wheel and axle machine, a larger wheel makes work easier than a small wheel.

6. **Comparison:** The conclusion and hypothesis agree.

7. **Presentation:** Prepare a presentation or a report to explain your results. If possible, set up the materials so that other people can try the experiment.

8. **Resources:** Tell the books you used to find background information. Tell who helped you get the materials and set up the experiment.

Other Project Ideas

1. How do the brakes on a bike work? Do research on friction and brake surfaces.

2. Some shapes are stronger than others. If you were building a bridge, which shapes would you use to support it? Make three-dimensional shapes from paper, and experiment to see how many books each shape can hold.

Unit 3: The Energy of Life

INTRODUCTION

The world is quickly changing. The use of energy has made these changes and advances in technology possible. This unit will explore the many kinds of energy that exist, including water, solar, wind, fuel, electric, nuclear, and magnetic. Students will even become aware of the limited resources that exist and look for ways to conserve energy.

SOURCES OF ENERGY

Energy is the ability to move things. We can see that the wind has energy when it blows a tree and that water has energy when it moves a boat. The force of the wind or water is being transferred to the object being moved. The more force exerted, the faster and farther it will go. The object, thus, has energy. The Sun and fossil fuels also have the ability to transfer force. Technology has found ways to harness all these different kinds of energy so that they produce electricity. Electricity is the energy that flows through wires. Power plants that produce electricity often use a variety of energy sources, such as water or fuels, to produce the electricity.

Once the energy has been transferred, an electric current has been made. An electric current is the movement of energy through a wire or a metal path. The current can then be transferred into heat, sound, or light. It can also cause movement or even make a magnet.

Water

Moving water has energy. In grain mills long ago, people discovered that the moving water transferred its energy to turn wheels. That energy then traveled to other wheels that would grind the grain. Modern technology has capitalized on this process and developed ways to utilize large amounts of water energy. Huge power plants use the movement of water to turn large wheels, called *turbines*, which run generators to produce electricity. The water energy is transferred to electric energy to produce electricity for houses.

Stored water has energy, too. As water is released from the elevated storage tank, the force of gravity pulls the water downward. The force of the stored water provides energy. A dam holds, or stores, water, but when released, the water pours through the gates with great force. The force of the water is transferred to anything that is in its path.

Wind

Wind is another source of energy. Wind currents are the result of uneven heating of the Earth's surface by the Sun. The air over the warm areas is heated. It expands, becoming less dense, and rises. Likewise, air over colder areas cools and becomes more dense. The cold air is heavier and sinks, allowing the lighter, warm air to rise. The movements of the cold and warm air produce wind. This wind has the energy to move leaves, trees, and in cases of storms, houses.

Windmills have been used for hundreds of years to use the wind's energy. Just like water, wind was used to grind grain. People also used windmills to pump water. As the wind turned the blades of the windmill, wheels in the device turned and transferred the energy into electric energy. Again, technology has improved the process by finding ways to transfer wind energy to electric power in larger capacities. Windmill farms in California and New Mexico collect the wind's energy for use in power plants.

Sun

The Sun provides light and heat. The heat energy of the Sun can be harvested to heat water and air in a house. This process is called *solar energy* and is a relatively new technology still being modified and improved. One of the simpler methods of solar energy utilizes the Sun's heat to shine through windows onto masonry floors of the house. The Sun's heat is transferred to the stone and slowly released to heat the room.

Another kind of solar energy involves the use of solar panels placed on the roof of a house. There are water pipes beneath the dark surface of the panels. The heat energy from the Sun is transferred to the water in the pipes. The pipes carry the heated water to faucets or to a storage tank built to hold the warm water. In some cases, the heated water passes a fan. The water heats the air, and the fan blows it into the house.

Heat from the Sun can be measured by a thermometer, a tool that measures how hot or cold something is. The two most common temperature scales are the Celsius scale and the Fahrenheit scale. Most scientific measurements are based on the Celsius scale. The freezing point, the point at which water freezes, is 0° on the Celsius scale and 32° on the Fahrenheit scale. The boiling point, the point at which water boils, is 100° on the Celsius scale and 212° on the Fahrenheit scale.

Candles

Candles provide heat and light. An outside source of energy needs to be transferred to the wick before the candle will produce its own energy. As the wax melts, a chemical change takes place. The solid wax is altered into a gas. Though the energy produced is small, it can be useful.

Fuels

Fuel is something that can be burned, such as gas, coal, and oil. When fuel is burned, heat is produced. The heat makes energy. Most fuels are fossil fuels and are found deep in the Earth. They were made from plants that lived millions of years ago. Harvesting the fuel from the ground is costly and dangerous. It then has to be refined before it can be used. There is also a limited supply of fossil fuel. As consumers continue to demand more energy, more fuel is needed.

Educated consumers can find ways to use less energy. Transportation and electricity take the largest amount of fuel resources. People can look for alternative modes of transportation that involve mass transportation, carpooling, or bike riding. Turning appliances and lights off when they are not in use is another way to use less energy. Also, recycling materials means less energy is used to make new products.

Nuclear Power

Nuclear power is a recent technological development. Parts of an atom are split, which produces an enormous amount of energy, called *nuclear energy*. The heat is transferred to water to make steam. Steam is then used to make electricity. There is no burning, so the process does not pollute the air. However, the process requires an element called *uranium*. Like fossil fuels, there is a limited supply on Earth. It is also dangerous, because radiation is produced. Any waste left from the nuclear energy has radiation, too.

ELECTRICITY

Electricity gives us light, sound, heat, and movement. It is caused by matter that has an electric charge—a negative charge. In electricity the matter is positively charged and moves in the same direction. By moving in one direction, the charge makes a current. The rate that the current flows depends on the number of charges and how much the wire resists the current.

Most electricity is made by generators. Generators push the charged particles in the same direction through a conductor. A conductor is a material that charged particles can move through easily. Wires are the main source of movement. Any material that a current cannot pass though, such as rubber, is an insulator. Rubber generally covers wires to keep the charges moving in one direction. Without the insulator, wires would heat and the moving energy would be lost.

Water is another good conductor of electricity. All electrical appliances need to be kept away from water, including rain. An electric shock is the result when the water and electricity interact. The contact could result in your immediate death. Electricity is an important part of life, but it can be dangerous if you do not use it carefully.

Other safety precautions should be taken when you are around electric devices. All outlets carry electricity. Anything put into a receptacle can transfer the electrical current to the object. Appliances that are not in use should be unplugged. Only plugs of mechanically safe appliances should be plugged into an outlet. Empty outlets should be covered with safety caps. Also, standing under trees in an electrical storm is dangerous. Since the body is a good conductor, a lightning strike in a tree could jump to a person's body.

Static Electricity

Matter is made of tiny particles. These particles have different charges. The charges can be positive or negative. If a particle has the same number of positive and negative charges, there is no charge—it is called *neutral*. When a neutral object gains or loses charges, static electricity is created. The charges move in all directions. Objects that have static electricity attract objects that have opposite charges. They repulse objects that have the same kind of charge. Also, light objects that have a neutral charge are attracted to objects with static electricity. Sometimes, a plastic comb will create static electricity as the hair is brushed. Since each hair is similarly charged, it repulses the other strands of hair.

Circuits

A circuit is the path an electric current travels. The charge leaves the energy source and moves through a wire to an end source that uses the electric charge, such as a light bulb or toaster. It then must follow a separate path back to its source. If the charge does not return to its original source, the charge builds up, causing the circuit not to work. When the circuit is completed, it is called a *closed circuit*. An open circuit is one in which a part of the path is missing. The charge is unable to follow its complete path. Most circuits have a switch so the flow of the current can be controlled.

Dry Cells

Dry cells, often used in energy experiments, store an electric energy. The inside is made of chemicals that create a chemical reaction with the mineral zinc. An electric current is formed. A dry cell has terminals at the top: one holds a positive charge and the other holds a negative charge. When wires are attached from each terminal to a source that uses an electric charge, a closed circuit is formed. Batteries are similar to dry cells, but can hold more charges than a dry cell.

MAGNETISM

Magnetism is a force that attracts metal materials, like iron, steel, nickel, and cobalt. The force is found in magnets, naturally found in the lodestone rock. They attract, or pull, and repel, or push, other pieces of metal. Synthetic magnets are made from steel or a combination of aluminum, nickel, cobalt, and iron. It is easy to transfer a magnetic charge to iron, but the charge will not last. Proper storage of synthetic magnets is important for them to retain their force.

Magnets come in all shapes and sizes. The force is focused at the end, or poles, of the magnets. These poles have a north and a south side. Most magnets are marked with an N and an S to identify the poles. (However, if they are not marked on a bar magnet, hang the magnet from a string. The north end of the magnet will point toward the north.) Like ends of two magnets repel each other. In other words, if two north ends of magnets were held together, they would repel each other. Unlike ends, a south and a north end, would attract each other. The area between the poles has some magnetic force, too, but it is not as strong as the poles.

Magnets do not need to touch, though. There is a magnetic force around each magnet called a magnetic field. When a piece of metal comes within a certain distance of the magnet, the magnet's field starts to pull the metal. The pull increases as the metal gets closer to the magnet. The size of the magnet affects the strength of the magnetic field.

Electromagnets

As electric charges move through a wire, a magnetic field is created around the wire. It is this force that is used to make an electromagnet. A metal bar is wrapped in wire and connected to an electric source, such as a dry cell. The more wire used, the greater the magnetic field and the stronger the magnet is. Doorbells and cranes in junkyards use the energy of electromagnets.

Name _____ Date _____

Blowing in the Wind

A. Do you know what things wind can move? Take a walk on a windy day. List the things you see that are moved by wind.

_____

_____

_____

B. Sometimes wind does useful things. Sometimes wind does things that people do not like. What are some of these things? Write your answers in the chart.

The Effect of Wind

Ways That Wind Helps People	Ways That Wind Bothers People

Unit 3: The Energy of Life

Name _____   Date _____

Make a Dam

A dam holds back water. When the water moves, it has energy. Energy is the ability to move things. Moving water has energy. It can turn a small pinwheel. It can also turn large wheels in machines that produce electric energy. The energy of water moving in a dam can be changed to electric energy.

✓ **You will need:**
clean milk carton   scissors
water   pinwheel

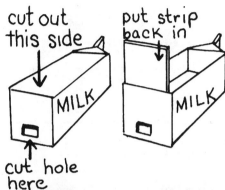

1. Cut off one side of the milk carton. Keep the piece in one large strip.

2. Cut a hole in the bottom of the carton.

3. Place the large strip behind the hole. Press it so it is snug. You will need to hold it in place.

4. Have a classmate help you. Hold the milk carton over the sink. Fill it with water. Press against the large strip to keep in the water.

5. Have your friend hold the pinwheel under the dam. Does the pinwheel turn? _____

6. Now pull out the strip. What happens?
   _____
   _____

Unit 3: The Energy of Life

Static Electricity

All matter contains charges. Charges can be rubbed off objects. When an object gains or loses a charge, it has static electricity.

✓ **You will need:**
paper
pencil shavings
plastic metric ruler
15 cm x 15 cm wool cloth
tape

1. Place the shavings on the paper.

2. Rub the wool cloth back and forth on the ruler about ten times. Hold the ruler over the pencil shavings. What happens?

3. Cut two pieces of tape, each about 15 centimeters long. Stick them to the edge of your desk so they hang off a little. What will happen if you move them close together?

4. Pull the tape pieces off the desk and hold them near each other. Was your prediction correct?

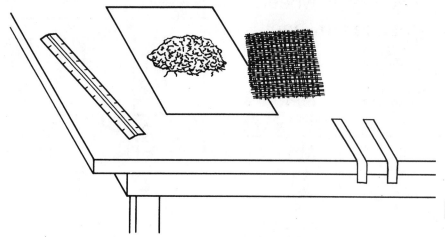

GO ON TO THE NEXT PAGE →

Name _____ Date _____

Static Electricity, p. 2

Answer the questions.

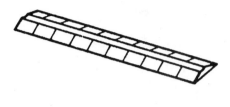

1. What happened when you held the ruler over the pencil shavings?

   _____
   _____
   _____

2. Why did you need to rub the ruler with wool?

   _____
   _____
   _____
   _____

3. What happened when you brought the two pieces of tape near each other?

   _____
   _____

4. What caused the two pieces of tape to act the way they did?

   _____
   _____
   _____
   _____

Name _____   Date _____

What Is Electric Current?

An electric current is made when charges move from place to place. Charges follow a path called a circuit. A dry cell or battery provides the charge. All parts of the circuit must be joined to work. A circuit that works is called a closed circuit. If any part is not joined, then the circuit is called an open circuit. In each of the pictures below, draw in the part needed to make a complete closed circuit.

1.

2.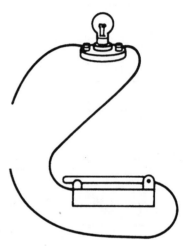

3.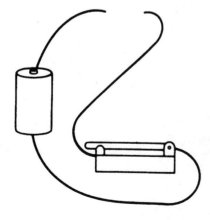

4.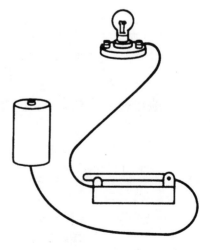

Unit 3: The Energy of Life

Name _____   Date _____

What Materials Conduct Electricity?

Electric currents can move through some materials more easily than others. Material that a current can move through is called a conductor. Most metals conduct electricity. They are like a wire that a current moves through. Your body is also a good conductor. Every circuit needs a generator, a conductor, and an electrical user.

You will need:
- 3 lengths of wire, each 30 centimeters long
- D-size battery
- toothpick
- key
- clear tape
- bulb and socket

NOTE: This experiment must be done with an adult.

1. Tape the end of a bare wire to the bottom of the battery.

2. Wrap the other end of the wire tightly around the socket of the bulb.

3. Tape another end of a bare wire to the top of the battery. Then tape the other end to the key.

4. Wrap the end of the third wire around the metal part of the bulb.

5. Touch the third wire to the key to make a closed circuit. What happens?

6. Replace the key with a toothpick. Repeat. What happens?

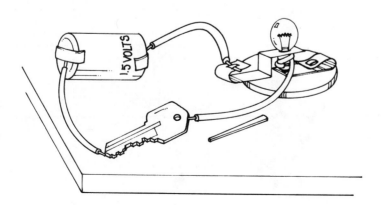

GO ON TO THE NEXT PAGE ➡

Unit 3: The Energy of Life

Name _____ Date _____

What Materials Conduct Electricity?, p. 2

Answer the questions.

1. Was the key a conductor?
   _____
   _____

2. Was the toothpick a conductor?
   _____
   _____

3. What is needed to make an electric circuit?
   _____
   _____
   _____
   _____

4. What kind of material is a good conductor?
   _____
   _____
   _____
   _____

Name _____   Date _____

Lemon Power

How are a lemon and a dry cell the same? You can get electric energy from both. Like a dry cell, a lemon has chemicals. You can change the chemical energy of the lemon to electric energy.

✓ **You will need:**
lemon   2 pieces of covered wire
large iron nail   battery-operated clock
scissors   orange, apple, and potato (optional)

NOTE: This experiment must be done with an adult.

1. Press hard on the lemon and roll it on your desk. This makes it juicy inside.

2. Cut off some of the covering from both ends of the wire.

3. Tightly wrap the end of one wire around the nail head. Push the nail halfway into the lemon.

4. Push one end of the second wire into the lemon. Make sure all of the uncovered part of the wire is pushed in.

5. Test your lemon battery to see if it makes electric energy. Place a clock next to the lemon.

6. Ask your teacher to touch the loose ends of the two wires together. Do the clock hands move?

7. See if you can make dry cells from other fruits and vegetables. Try an orange, an apple, and a potato.

Unit 3: The Energy of Life
Physical Science 3, SV 3762-3

Name _____ Date _____

Inside a Light Bulb

Find out what happens when electric current runs through the thin wires in a bulb.

You will need:
lamp   light bulb   dry cell   thin copper wire   oven mitt

NOTE: This experiment must be done with an adult.

1. Observe a clear light bulb. What do you see inside it?

2. Have your teacher put the bulb inside a lamp. See what happens when the lamp is turned on.

3. Have your teacher attach one end of the wire to one of the poles of the dry cell.

4. The wire may get hot, so your teacher should wear an oven mitt. Have your teacher wrap the wire once around the other pole of the dry cell and pull the wire tight.

Answer the questions.

1. What made the light in the light bulb? _____

2. What happens to the wire when joined to the dry cell?

3. Explain what happens to the wire in the light bulb.

Unit 3: The Energy of Life

Using Electricity

Electrical energy can be changed into different forms of energy. Appliances we use every day help the energy to change.

Match each object shown at the left with the word in the right-hand column that tells what it changes electricity into.

1.  Light bulb   movement

2. 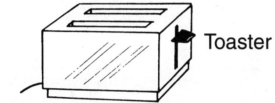 Toaster   sound

3.  Electric fan   heat

4. Doorbell   light

Name _____ Date _____

How Can You Reduce Electrical Use in Your School?

Most electricity is made from fuel that can be burned, such as oil, gas, or coal. Some electricity is made from the energy of the Sun, water, and wind. Some energy is made from a process called nuclear energy where atoms are split. These ways of making energy can cost a lot. Moreover, some of the fuels are in short supply. To save fuel, people need to find ways to save energy.

Look around your school or home. Find ways that electricity is used. Write four ways on the table below.

Complete the table.

Electrical Use

Object	Drawing of Object	What Object Does	Who Uses Object	How Object Can Be Used Less Often

Write three reasons it is important to reduce electrical use.

_____

_____

_____

Unit 3: The Energy of Life

Name _____ Date _____

Safe or Not Safe?

The pictures show ways to use electric energy. If a picture shows a way that is safe, write **safe** under it. If a picture shows a way that is not safe, write **not safe** under it.

1.
_____

2.
_____

3.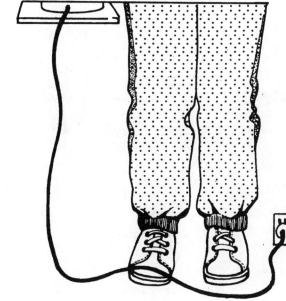
_____

4.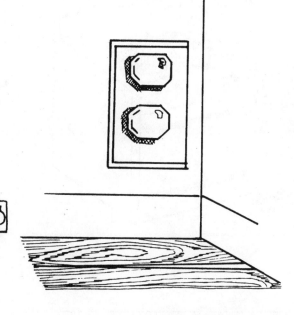
_____

GO ON TO THE NEXT PAGE ➤

Unit 3: The Energy of Life

Name _____ Date _____

Safe or Not Safe?, p. 2

5.

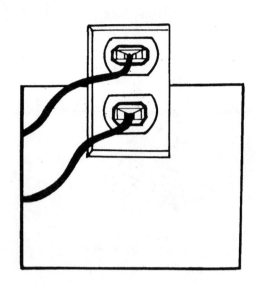

_____

6.

_____

7.

_____

8.

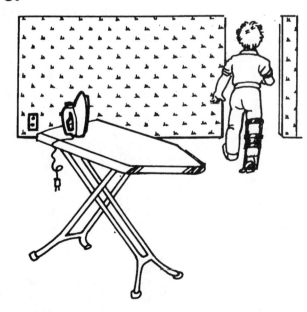

_____

Unit 3: The Energy of Life

Name _____ Date _____

Baked Apple—The Solar Way

The Sun makes heat energy. The energy can be used to dry clothes. The Sun's energy also can be used to cook food. Use the Sun's energy to cook an apple.

✓ **You will need:**
- 2 pieces of stiff paper
- sheet of black paper
- plastic food wrap
- aluminum foil
- paper cups
- apple slices
- brown sugar
- cinnamon
- clay
- tape
- scissors
- clock

1. Tape two pieces of stiff paper together end to end. Cover one side of the paper with aluminum foil, shiny side up. Tape the foil in place.

2. Now line the inside of the paper cup with black paper. Tape the paper in place. Do not put paper in the bottom of the cup.

3. Put two thin slices of apple in the cup. Sprinkle brown sugar and cinnamon on the slices.

4. Cover the top of the cup with plastic food wrap. Tape it down.

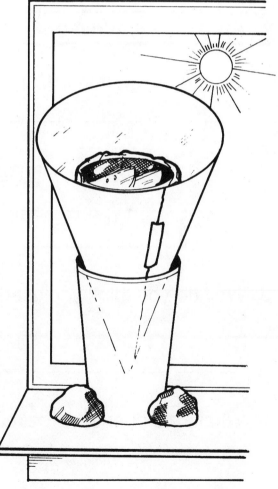

GO ON TO THE NEXT PAGE ➤

Unit 3: The Energy of Life

Baked Apple—The Solar Way, p. 2

5. Bend the heavy paper into a cone around the cup. Make sure the foil is on the inside. Tape the cone in place.

6. Place the cone and the cup inside another paper cup. Place your solar heater in a sunny spot. Hold it in place with clay. Look at the slices after three hours.

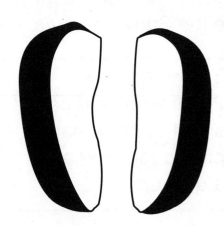

 Answer the questions.

1. What happened to the apple slices? _____
   _____

2. How does the aluminum foil help the process?
   _____
   _____

3. Why do you think you used plastic wrap over the apples?
   _____
   _____

4. What kind of energy "baked" the apples?
   _____

Does the Heat from the Sun Affect Everything in the Same Way?

The Sun makes heat energy. People can use the Sun to heat the water and air in their homes by using solar energy. Dark panels, called solar panels, are placed on the roof. There are pipes filled with water in the solar panels. The energy from the Sun heats the water in the pipes. The heated water moves to the pipes of the sink and shower in the house. The water can also heat the air in the house.

✓ You will need:
newspaper
black construction paper
white construction paper
clock

1. Unfold several newspapers and lay them on a windowsill in the sunlight.

2. Put the white and black paper in the sunlight side by side.

3. Wait 15 minutes. Touch each paper. Which paper felt warmer?

Answer the questions.

1. Which paper felt warmer? _____

2. Which paper felt cooler? _____

3. Which color holds the heat better? _____

4. Why do you think solar panels are painted black? _____

_____

Unit 3: The Energy of Life

Name _____   Date _____

Measuring Heat

A thermometer is a tool that measures how hot or cold something is. A thermometer usually shows two scales: Fahrenheit and Celsius. Most scientists use the Celsius scale. The top of the thermometer has *C* if it is a Celsius scale. We usually read the Fahrenheit scale. The thermometer will show this scale with an *F*. The thermometer shows the temperature in units called degrees. The degrees are shown with lines and numbers. The higher the red line is on the thermometer, the warmer the temperature is.

The thermometer shows temperatures using the Fahrenheit scale.

Write each temperature.

1.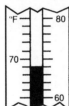

2.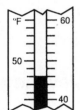

3.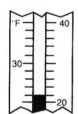

_____   _____   _____

Draw a line to match the temperature on each thermometer to the clothes that should be worn.

4.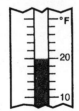

5.

6.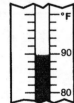

Unit 3: The Energy of Life

Name _____ Date _____

How Can You Measure Temperature?

On a Celsius scale, 100° is the boiling point, the point at which water boils, and 0° is the freezing point, the point at which water turns to ice. On a Fahrenheit scale, 212° is the boiling point, and 32° is the freezing point. To read a thermometer, look at the end of the red or blue line. Then look at the degrees, the numbers that show the scale. You may have to estimate the temperature by rounding the number of degrees to the nearest number.

✓ **You will need:**
Styrofoam cup	ice cubes	spoon
Celsius thermometer	warm water	clock

NOTE: This experiment must be done with an adult.

1. Fill the cup with warm water.

2. Put the thermometer into the water. When the red or blue line stops moving, read the temperature. Record the temperature on the table on the next page.

3. Remove the thermometer.

4. Put ice cubes into the warm water. Stir the water for one minute.

5. Repeat step 2.

GO ON TO THE NEXT PAGE ➤

Unit 3: The Energy of Life

Name _____ Date _____

How Can You Measure Temperature?, p. 2

Water Temperature

Water	Temperature (C°)
Warm	
Cold	

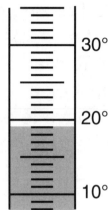

 Answer the questions.

1. What was the temperature reading of the warm water?
   _____

2. What happened to the liquid in the thermometer after you put the ice cubes in the water?
   _____
   _____

3. Did the ice make the water cooler? How do you know?
   _____
   _____

4. When would you need to read a thermometer?
   _____
   _____
   _____

Unit 3: The Energy of Life

Name _____ Date _____

Measuring Heat

In each of these thermometers, the red liquid is missing. There is a temperature reading under each picture. Color in the temperature for each picture.

1.
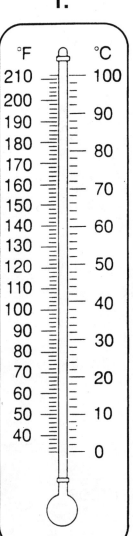
70°C

2.
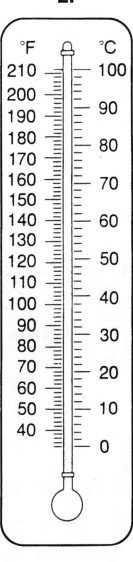
Boiling point of water

3.
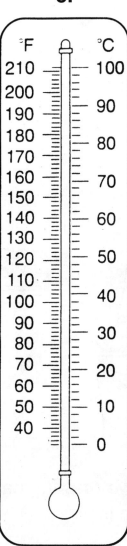
Freezing point of water

4.
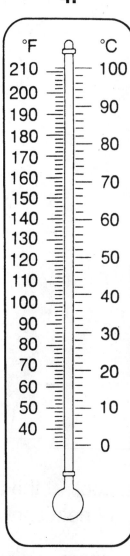
Normal body temperature

Name _____ Date _____

Sources of Heat

Look at the above drawing carefully. Then list the sources of heat shown in it.

_____
_____
_____
_____

Unit 3: The Energy of Life

Popping Corn with a Candle

A candle is usually used as a light source. You can use its energy for heat, too. You can use a candle to make popcorn! So put away the fancy popcorn poppers and try candle power instead.

You will need:
- 10 popcorn kernels
- aluminum foil
- ring stand
- paper clip
- matches
- candle in dish
- matches
- clock

NOTE: This experiment must be done with an adult.

1. Place the popcorn on the square of aluminum foil.

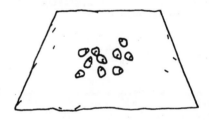

2. Twist the corners of the foil together to make a bag. Be sure to leave a lot of extra room. The popcorn gets much bigger when it pops.

3. Straighten out the middle bend in the paper clip. Hook one end through the bag. Hang the other end over the crossbar of the ring stand.

GO ON TO THE NEXT PAGE

Unit 3: The Energy of Life

Name _____ Date _____

Popping Corn with a Candle, p. 2

4. Have an adult light the candle and put it under the foil bag.

5. Stand back. Watch—and listen to—what happens!

6. Wait 15 minutes for the popcorn to cool. Enjoy!

Answer the questions.

1. What did you observe? _____
_____

2. What energy did the candle give? _____
_____
_____

3. What part of the candle made the energy? _____
_____
_____

4. What did the energy of the candle wax do? _____
_____
_____
_____

© Steck-Vaughn Company

Unit 3: The Energy of Life
Physical Science 3, SV 3762-3

Name _____ Date _____

Energy Detective

A. Be an energy detective. Find out what kinds of energy are used in your school, home, and neighborhood. Find out how the energy is being used. Look for electric, chemical, solar, and wind energy.

Fill in the chart.

Energy Uses

Kind of Energy	How It Is Used
example: chemical	candle—gives off light

B. See if you can find any places in your home where energy is sometimes wasted. Can you think of ways to stop the waste of energy?

Fill in the chart.

Where Energy Is Wasted

Place	How to Stop Energy Waste
1. Example: the television is on when no one is watching.	Turn the television off when you are finished watching it.

Unit 3: The Energy of Life

Name _____   Date _____

Oil Traps

Long ago, tiny plants and animals lived in the ocean. When they died, they sank to the ocean floor. As years passed, layers of mud and sand covered them. Slowly, the mud and sand changed to rock. Sometimes the mud became shale. The sand turned to sandstone. The dead plants and animals became tiny drops of oil.

The drops of oil moved away from some places and collected in others. Find out why oil was trapped in some places and not in others.

✓ **You will need:**
vegetable oil   eyedropper   ball of clay
spoonful of sand   wax paper
different kinds of rocks (including sandstone and shale)

1. Place the sand on a piece of wax paper. Using the dropper, place three drops of oil on top of the sand. Does the oil stay on the top of the sand or does it sink into the sand? _____

2. Place the ball of clay on another piece of wax paper. Flatten the clay. Place three drops of oil on top of it. What happens to the oil?

_____

3. Place the different kinds of rocks on wax paper. Test each one in the same way. What do you find? _____

_____

_____

© Steck-Vaughn Company

Unit 3: The Energy of Life
Physical Science 3, SV 3762-3

Using Crude Oil

Read the story about petroleum products.

When petroleum comes from the ground, it is called crude oil. Crude oil is a mixture of many substances. These substances must be separated before they can be used.

Crude oil is first taken to a refinery. It is heated to 400° C in a special furnace. At this temperature the crude oil boils. Most of it becomes a mixture of gases. A pipe brings the gases to the bottom of a tower. The gases rise through the tower. As they rise, the gases cool. Most become liquid again. The liquids are collected and stored in tanks.

Different petroleum products become liquid at different levels in the tower. For example, gasoline turns to liquid near the top. Heating oil turns to liquid near the middle. A few gases do not get cooled enough to change back to liquids. These are drawn off at the top of the tower. Some material is left at the bottom of the tower. This part of the crude oil did not change to a gas. Asphalt is the main product in this material. It is used to build roads. Look at the picture on the next page to see where other products are separated in the tower.

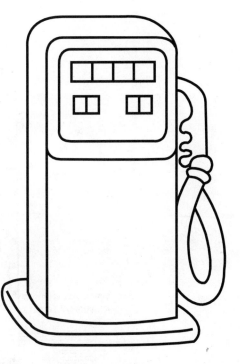

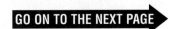

GO ON TO THE NEXT PAGE

Name _____ Date _____

Using Crude Oil, p. 2

Put the sentences in the right order. Number them from 1 to 6.

_____ Crude oil is taken to a refinery.

_____ The gases turn to liquid at different heights in the tower.

_____ Different petroleum products are collected at different levels of the tower.

_____ Crude oil comes out of the ground.

_____ In a furnace, crude oil is heated to 400° C.

_____ Crude oil changes to gases and rises in the tower.

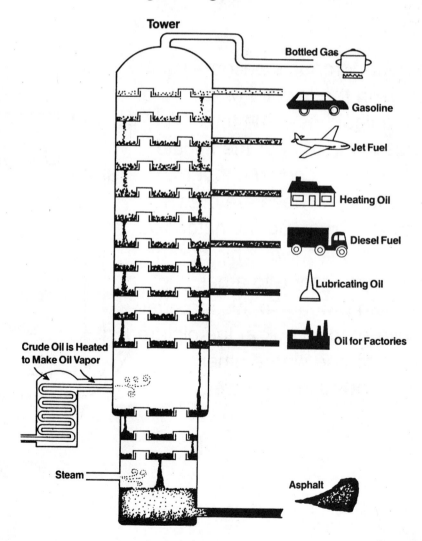

Unit 3: The Energy of Life

Name _____   Date _____

Petroleum Around the World

Petroleum has many uses. Gasoline, heating fuel, plastics, detergents, and drugs are just a few things that are made from petroleum.

Most countries do not produce as much petroleum as they use. So they must import, or bring in, the petroleum they need from other countries. Look at the table below. It shows how much petroleum is used in these places.

Petroleum in the World

Place	Petroleum Produced	Petroleum Used
Middle East	🛢🛢🛢🛢🛢	🛢🛢🛢
Europe and Russia	🛢🛢🛢🛢	🛢🛢🛢🛢
USA and Canada	🛢🛢	🛢🛢🛢🛢🛢
Latin America	🛢🛢🛢🛢	🛢🛢🛢
Asia	🛢🛢🛢	🛢🛢🛢🛢
Africa	🛢🛢	🛢🛢

Answer the questions. Use the table to help you.

1. Where is most of the petroleum produced? _____

2. What place uses as much petroleum as it produces?

   _____

GO ON TO THE NEXT PAGE ▶

Unit 3: The Energy of Life

Petroleum Around the World, p. 2

3. What places must bring in most of the petroleum they use? _____

4. The United States uses more petroleum than it produces. Where does it get the rest of its needs? _____

5. What place produces more petroleum than it uses? _____

6. Should people use less petroleum? Explain. _____

7. What can people do to save petroleum? _____

What Can a Magnet Pick Up?

Magnets pick up and stick to many different things. They come in different shapes and sizes. They have different strengths, depending on their size. They have a force called a magnetic force. If not stored properly, magnets can lose their magnetic force.

✓ **You will need:**
magnet	paper clips	pencil
screw	washer	eraser
plastic protractor or ruler		penny

1. Place all the objects except the magnet on a table.

2. Predict what the magnet will pick up. Record your predictions on the table on the next page.

3. Use the magnet to try to pick up each object. Which objects did the magnet pick up? Record your findings on the table.

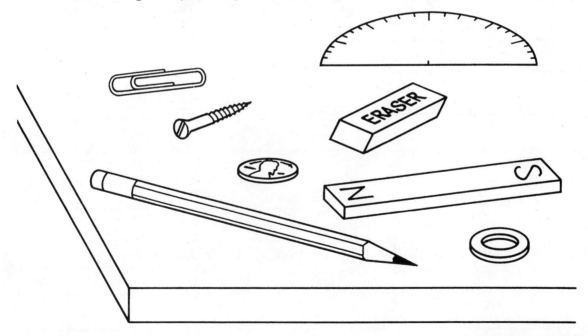

GO ON TO THE NEXT PAGE

Unit 3: The Energy of Life

Name _____ Date _____

What Can a Magnet Pick Up?, p. 2

What Magnets Pick Up

Object	Prediction	Result

 Answer the questions.

1. Which objects stuck to the magnet? _____
   _____

2. What are these objects made of? _____

3. Name three other objects that a magnet can pick up.
   _____
   _____

Unit 3: The Energy of Life

Name _____ Date _____

How Do Poles of a Magnet Act?

Each end of a magnet is called a pole. There is a north pole, labeled N, and a south pole, labeled S. The magnetic force is strongest at the poles. The poles have different forces. If you put the ends of the poles of two magnets close to each other, the magnets will either push away from each other or pull toward each other. If the magnets pull together, the poles are unlike poles. If they push apart, the poles are like poles.

You will need:
2 bar magnets   3 books
metric ruler   30-centimeter piece of string

1. Tie one end of the string to the center of the bar magnet. Tie the other end to the end of the ruler.

2. Place the books in a stack near the edge of the desk. Place the ruler between the books. Stop the magnet from moving.

3. Bring the north pole of the second magnet near the north pole of the hanging magnet. What happens?

4. Bring the north pole of the second magnet near the south pole of the hanging magnet. What happens?

5. Repeat steps 3 and 4 using the south pole of the second magnet. What happens?

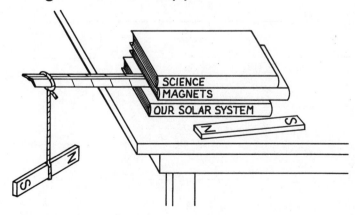

GO ON TO THE NEXT PAGE ▶

Unit 3: The Energy of Life

Name _____ Date _____

How Do Poles of a Magnet Act?, p. 2

Answer the questions.

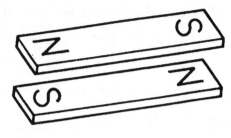

1. What happened when the north poles were near each other?
   _____
   _____

2. What happened when the south poles were near each other?
   _____
   _____

3. What happened when a north pole was near a south pole?
   _____
   _____

4. How do like poles act?
   _____
   _____

5. How do unlike poles act?
   _____
   _____

Unit 3: The Energy of Life

Name _____ Date _____

A Magnet and an Electromagnet

An electromagnet is a magnet made from a wire nail and dry cell. The current of the dry cell causes the nail to have a magnetic force.

Does an electromagnet work as well as a magnet?

You will need:
electromagnet   paper clips
magnet   stack of paper
objects to test such as buttons, coins, aluminum foil, rubber band, thumbtack, nail, and pencil

1. Write *PULLED* on one sheet of paper. Write *NOT PULLED* on another.

2. Test different objects with the electromagnet. Then put each object on top of the paper that describes what happened.

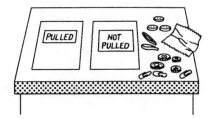

3. Test the objects again. Use the magnet this time. Was there any object that was pulled by the electromagnet but not by the magnet? Was there any object pulled by the magnet but not by the electromagnet?

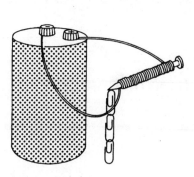

4. Now, find out which magnet can hold more. Line up some paper clips. See how many paper clips the electromagnet can hold.

GO ON TO THE NEXT PAGE

Unit 3: The Energy of Life

A Magnet and an Electromagnet, p. 2

5. Use the magnet to make a line of paper clips.

6. Place a sheet of paper over a paper clip. Can the electromagnet still pick up the clip? Find the greatest number of sheets of paper it can pull through.

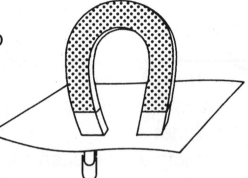

7. Repeat step 6 using a magnet. Find the greatest number of sheets of paper it can pull through.

 Answer the questions.

1. Did the electromagnet or magnet attract more objects?
_____

2. Did the electromagnet or magnet hold more paper clips in a line?
_____

3. Which is stronger—the electromagnet or the magnet?
_____

Name _____ Date _____

Unit 3 Science Fair Ideas

A science fair project can help you to understand the world around you. Choose a topic that interests you. Then use the scientific method to develop your project. Here's an example:

1. **Problem:** What part of a magnet holds the most?

2. **Hypothesis:** The poles of a magnet have more magnetic force.

3. **Experimentation:** Materials: different shaped magnets, paper clips, centimeter ruler
   - Find out how many paper clips the poles of each magnet will hold.
   - Find out how many paper clips the center of each magnet will hold.
   - Measure the strength of each magnet. Place a paper clip at one end of the ruler and the end of a magnet at the other. Slowly slide the magnet closer to the clip. When the paper clip starts to move, read the distance it is from the magnet.
   - Repeat the above step using the center of the magnets.

4. **Observation:** The poles of the magnets attracted more paper clips. The end of the magnet was also a greater distance from the paper clip before it moved.

5. **Conclusion:** The poles of the magnet have the greater magnetic force.

6. **Comparison:** The conclusion and the hypothesis agree.

7. **Presentation:** Prepare a presentation or a report to explain your results. If possible, set up the materials so that other people can try the experiment.

8. **Resources:** Tell the books you used to find background information. Tell who helped you to get the materials and set up the experiment.

Other Project Ideas

1. What color clothing should people wear in the summer? Use thermometers and colored fabric to experiment.

2. How do people use wind to make energy? Do research on windmills and plan a project.

Physical Science Grade 3
Answer Key

p. 9 1. a. 2. d. 3. a. 4. b. 5. a.
p. 10 6. a. 7. b. 8. c. 9. d.
p. 11 1. b. 2. a. 3. b. 4. b. 5. c. 6. c
p. 12 1. a. 2. c. 3. b. 4. c. 5. b.
p. 13 6. b. 7. a. 8. c. 9. b. 10. b.
p. 18 1. matter 2. No; Since everything is different, it will not be described the same way.
p. 19 1. matter 2. You find out about matter by using your senses.
p. 21

STUDYING MATTER

	Solid	Liquid	Gas
Does it take up space?	yes	yes	yes
Does it have shape of its own?	yes	no	no
Does its shape depend on the shape of the container?	no	yes	yes
Can it be seen?	yes	yes	sometimes
Does it always stay the same size?	yes	yes	no
Does it spread out to fill up its container?	no	no	yes

1. solid, liquid, gas 2. Answers will vary.
3. Possible answers: solids: skin, bones, hair, organs; liquids: blood, saliva; gas: oxygen in lungs inhaled, carbon dioxide exhaled.
p. 22 1. solid 2. gas 3. liquid 4. gas 5. liquid 6. solid
p. 23 1. Answers will vary.
p. 24 1. If the balance is weighted in one direction, the measurement of mass will not be accurate. 2. Find the sum of the gram measures. 3. Answers will vary.
p. 25 1. liquid; Liquids take the shape of a container. 2. beaker 3. Answers will vary.
p. 26 1. The unpopped balloon sinks. 2. It is now heavier than the popped balloon. 3. The air inside the unpopped balloon adds weight to it.
p. 27 1. smell; air 2. see; liquid 3. smell; air 4. see; liquid
p. 28 1. Answers will vary. 2. Answers will vary. 3. Molecules of substances travel through the air. I can smell the substances when the molecules reach my nose.
p. 29 1. The balloon is smaller. The air molecules were able to pass through the balloon. 2. It took longer in cold water. Molecules move faster in warm liquids than in cold.
p. 30 1. true 2. false 3. true 4. false 5. false
p. 31 1. Answers will vary. 2. Answers will vary, but should suggest that the place they put the ice cube was warmer; therefore, heat made the ice cube melt. 3. Answers will vary.
p. 32 1. The towel on the windowsill dried first. 2. It was hotter on the windowsill, so the water evaporated faster. 3. The water evaporated into the air.
p. 34 1. pie tin; Answers will vary. 2. soda bottle; Answers will vary. 3. The container with the smallest opening took the longest to evaporate. The container with the largest opening took the least time to evaporate.

p. 35 1. water droplets 2. clear 3. no 4. They would be colored if they came from inside the can. 5. Possible answers: on beverage glass, on mirrors in bathrooms after a shower, on windows, on grass, and on outdoor furniture.
p. 36 2. Answers will vary. 3. Answers will vary; 1. no 2. All the matter stayed in the balloon. No water was taken out or added. Since no matter is lost or gained when the state changes, the weight does not change.
p. 37 1. Students draw an empty glass. 2. Students draw a bigger balloon. 3. Students draw small amount of liquid on plate.
p. 38 1. The ice cubes cool the glass. The glass cools the water vapor around the glass. 2. Water becomes ice. 3. The iron becomes a solid iron bar. 4. The air cools the water vapor rising from a boiling pot.
p. 40 1. solid and liquid 2. Heat was added to make the soap become a liquid and heat was removed for it to return to a solid. 3. Small soap pieces, ordinarily thrown out, were combined to make a large soap bar. 4. Answers will vary.
p. 41 1. 2, 1, 3 2. Students circle the picture in which the jar is being heated.
p. 42 1. a mixture 2. No; The states of matter did not change form. Their properties stayed the same. 3. Possible answers: Let water evaporate, filter water, or get marbles out by hand. 4. Answers will vary.
p. 43 1. *Dissolved* means that a solid is mixed with a liquid so that you cannot tell the two apart. 2. It dissolved in the hot water faster, because molecules move faster to help the sugar change states. 3. The solid can be separated from the liquid. 4. Let the water evaporate.
p. 44 1. It dissolved faster in the cup with the hot water and crumbs. 2. It dissolves faster. 3. The hot water makes the solid dissolve faster. 4. There were bouillon crystals.
p. 46 1. No 2. The middle and bottom of the cube turned color. 3. water 4. food color and water
p. 48 1. It climbed up the paper. 2. It separated into different colors. 3. Answers will vary depending on color of ink. 4. Check students' drawings. 5. no 6. Answers will vary.
p. 50 1. fuel, oxygen, heat 2. fuel-candle; oxygen-air; heat-matches 3. Answers will vary. 4. Answers will vary. 5. oxygen 6. Yes; A fire needs oxygen to burn.
p. 51 A. 1. c. 2. a. 3. b. **B.** 1. The candle would go out. 2. The rusting would stop. 3. The boy would not be able to live.
p. 52 1. Physical change; The molecules do not change. 2. Physical change; The molecules do not change. 3. Chemical change; The properties of the rubber band changed because it crumbles when touched and does not stretch.

Answer Key

p. 53 1. chemical 2. physical 3. physical
4. physical 5. chemical 6. chemical

p. 54 1. A chemical change took place when the flour, sugar, and margarine were heated together. 2. heat 3. A physical change took place. Even though heat was added, the raisins still look and taste the same. 4. A chemical change would have taken place, because the cookies would taste and feel different.

p. 56 1. A chemical change happened because the molecules in the 2 substances changed.
2. A physical change would have taken place. The vinegar would have been dissolved into the water to make a solution.

p. 64 1. pull 2. push 3. push 4. pull 5. pull 6. push

p. 65 1. b. 2. d. 3. c. 4. a. 5. 4 newtons; They would both have the same reading because the same amount of force is acting on both spring scales. 6. B. 7. A.

p. 66 Students color the last picture.

p. 67 1. The total force is 100 newtons + 75 newtons = 175 newtons. The wagon will move forward. 2. The wagon will move backward, because Sara is applying a greater force than David is.

p. 68 3. The rope will move toward Emily's team. 4. 260 newtons – 60 newtons = 200 newtons; The rope will move toward Chris's side, because his team is pulling with greater force. 5. The boxcar will not move, because the two forces are equal, or balanced. 6. The total force on the boxcar is 2,000 newtons. The boxcar will move to the right.

p. 69 1. b. 2. d. 3. c. 4.

Mass on Earth	Weight on Earth	Mass on Moon	Weight on Moon
3 kilograms	30 newtons	3 kilograms	5 newtons
6 kilograms	60 newtons	6 kilograms	10 newtons
9 kilograms	90 newtons	9 kilograms	15 newtons
12 kilograms	120 newtons	12 kilograms	20 newtons
15 kilograms	150 newtons	15 kilograms	25 newtons

p. 70 Students color the floating swing, child, trashcan, dog, and umbrella.

p. 71 1. Students color girl walking uphill.
2. Students color picture of car moving uphill.

p. 72 1. brick 2. book 3. bowl of apples 4. barbell

p. 74 1. He wanted to prove that heavier objects fall faster. 2. He dropped 3 stones off the bridge and timed how long they took to hit the water.
3. 5 seconds 4. No matter the weight of the stones, they all took the same amount of time to fall.
5. Gravity pulls objects, no matter their mass, at the same rate of speed.

p. 76 Check students' tables. 1. As the number of washers increased, the cup took longer to slide across the table. 2. The cup would move more slowly. With enough washers, it might not move at all. 3. You should put something heavy in the basket.

p. 77 1. c. 2. Forces can work together, so the force of two people pushing in the same direction would be greater than the force of one person pushing.

p. 78 1. No; The marble appears to move in the opposite direction that the box moves. 2. Inertia; The marble stays still because it was still.

p. 79 1. grass 2. thick carpet 3. towel 4. gravel
5. a. reduce friction 6. There would be less friction between the box and the floor, so the box could be pushed more easily.

p. 80 1. The first way was easier, because friction between the hands and lid caused the lid to move.
2. soap and water

p. 82 1. Adam hypothesized that the sneakers would grip the floor better than the sandals and that the wood floor was more slippery than the carpet. 2. Sneaker: The data on the table shows the sandal slid 3 out of 4 times. 3. Carpet: The data on the table shows the shoes slid on the wood floor 3 out of 4 times. 4. By testing the shoes on each surface, Adam was able to find out which combination of shoes and floor surface would be best for his sister to walk on.

p. 83 1. Students color second picture. There is less friction on the wood floor than on the carpet.
2. Students color first picture. There is less friction on the smooth floor than on the grass.

p. 84 1. b. 2. It would roll toward the back of the bus, because there would not be anything to push against it to make it move with the bus. 3. It would roll toward the front of the bus, because there would not be anything to push against it and make it slow down with the bus.

p. 85 Answers are given in order from top to bottom: 4, 1, 3, 2

p. 86 Students circle 1, 3, 4.

p. 87 1. first-class lever

p. 88 2. second-class lever 3. third-class lever
4. The nut will not crack open. 5. The crowbar is the first-class lever. 6. Answers will vary. Possible answers include bottle opener, poptop of can, pronged end of hammer.

p. 89 1. the book 2. When the fulcrum is closer to the load, less force is needed to lift the load.

p. 90 1. The broom handles act like a wheel and the rope is wrapped around them to make a pulley.
2. It changes the direction of the force.

p. 92 1. It took more force to lift the book without using the ramp. 2. Using the ramp required moving the book a greater distance. 3. Less force was needed to move the book by using the ramp.
4. Friction could make it harder to push or pull an object up the ramp. You would need more force to get the object up the ramp. 5. Possible answers include: moving van, stairs, mountain road, wheel chair ramps.

Answer Key

p. 93 Students circle 1, 2, 4, 6.

p. 94 1. The thread of the screw is the inclined plane. 2. The point of the screw is a wedge made by 2 inclined planes joined together. 3. They are alike because inclined planes are used to move things with less force. They are different because a ramp is a straight inclined plane that stands still. A screw has a wrapped inclined plane that moves.

p. 96 1. The handle is the lever that moves the wheel and axle. 2. Answers will vary, but should be around 4 times. 3. The suds were more frothy in the bowl mixed with the eggbeater, because it used more force in the same amount of time. 4. Answers might include fishing pole, can opener, hand-crank ice cream machine, water well.

p. 98 1. Possible response: A rope attached to Jay's leg was strung through the pulley and attached to a weight. The pulley turned the downward force of the weight into an upward force to pull on the leg. 2. Possible response: There were 2 wheels and axles on the back of the wheelchair. The large handle helped Jay use the chair easily. 3. Possible response: The ramp Jay used was an inclined plane. He could go from one level to another with less force than moving his chair and himself straight up and down.

p. 99 1. Students label the blades as wedges and the handles as levers. 2. The fulcrum is the piece that joins the handles together so they open and close. 3. The force of the wedges helps to cut the paper in a straight line. 4. Answers will vary.

p. 106 Answers will vary.

p. 107 The moving water has energy, causing the pinwheel to turn.

p. 109 1. The shavings jumped toward the ruler. 2. The wool created an electric charge in the ruler. 3. The tape pieces moved away from each other. 4. They received the same kind of charge when they were pulled away from the desk. Like charges repulse each other.

p. 110 1. Students draw wires to all three parts. 2. Students draw the battery. 3. Students draw the bulb. 4. Students draw the wire from the battery to the bulb.

p. 112 1. yes 2. no 3. A generator, a conductor, and an electrical user are needed to make an electric circuit. 4. metal

p. 114 1. The wire makes the light. 2. The wire gets hot and gives off light. 3. Electric energy is moving through the wire. It gets hot and glows, making light in the bulb.

p. 115 1. light 2. heat 3. movement 4. sound

p. 116 Answers will vary.

p. 117 1. not safe 2. not safe 3. not safe 4. safe

p. 118 5. safe 6. not safe 7. not safe 8. safe

p. 120 1. The slices turned brown. They were cooked. 2. Possible response: The foil helps to make more heat. 3. The plastic wrap holds the heat in. 4. solar energy

p. 121 1. The black paper felt warmer. 2. The white paper felt cooler. 3. The black paper holds the heat beater. 4. Since black holds the heat, solar panels are painted black to hold the heat better.

p. 122 1. 68° 2. 46° 3. 22° 4. Students draw a line to the snowsuit. 5. Students a draw line to the coat and pants. 6. Students a draw line to the swimsuit.

p. 124 1. Answers will vary. 2. It moved down. 3. Yes; The temperature reading went down, showing the temperature of the water went down. 4. Possible answer: It would help me to decide what clothes to wear.

p. 125 1. Students color to 70°C. 2. Students color to 100°C. 3. Students color to 0°C. 4. Students color to about 98.6°F.

p. 126 sun, charcoal, wood, gas, electricity

p. 128 1. There were popping sounds. The foil got bigger. 2. heat and light 3. The wax makes the energy. 4. The energy heated the popcorn kernels. They swelled, popped, and pushed against the foil. The heat energy was moved to the popcorn.

p. 129 Answers will vary.

p. 130 1. It sinks in the sand. 2. The oil stays on top of the clay. 3. Answers will vary depending on the rocks used: Oil will flow through sandstone, but not through shale.

p. 131-132 Sentence order: 2, 5, 6, 1, 3, 4

p. 133 1. Middle East 2. Africa

p. 134 3. Asia, United States, Canada 4. It must import from other countries. 5. Middle East 6. Yes; There is a limited amount of petroleum. 7. possible answers: Drive only when necessary, take a bus or subway, ride a bike, carpool.

p. 136 1. The magnet will pick up the paper clip, screw, washer. 2. steel or iron 3. Answers will vary.

p. 138 1. The north pole of the hanging magnet moved away. 2. The south pole of the hanging magnet moved away. 3. The hanging magnet moved toward the magnet being held. 4. They push away from each other. 5. They pull toward each other.

p. 140 1. They attracted the same objects. 2. Answers will vary. 3. Answers will vary.

Answer Key